Auditory Physiology

Auditory Physiology

Aage R. Møller

Division of Physiological Acoustics
Department of Otolaryngology
University of Pittsburgh School of Medicine
Pittsburgh, Pennsylvania

ACADEMIC PRESS   1983
A Subsidiary of Harcourt Brace Jovanovich, Publishers
New York   London
Paris San Diego San Francisco São Paulo Sydney Tokyo Toronto

ACADEMIC PRESS, INC.
111 Fifth Avenue, New York, New York 10003

United Kingdom Edition published by
ACADEMIC PRESS, INC. (LONDON) LTD.
24/28 Oval Road, London NW1 7DX

Library of Congress Cataloging in Publication Data
Main entry under title:

Auditory physiology.

 Includes bibliographies and index.
 1. Hearing. 2. Auditory pathways. I. Møller,
Aage R. Date. [DNLM: 1. Acoustic nerve--
Physiology. 2. Ear--Physiology. WV 272 M693]
QP461.A928 1982 612'.85 82-8903
ISBN 0-12-503450-4

PRINTED IN THE UNITED STATES OF AMERICA

83 84 85 86   9 8 7 6 5 4 3 2 1

Contents

3 Frequency Analysis in the Peripheral Auditory System

4 Coding of Complex Sounds in the Auditory Nervous System

Preface

In man, hearing is the most important sense for communication between individuals. Our present society greatly relies on speech communication, which depends upon the intactness of hearing.

The efficiency of the auditory system, as well as of the entire nervous
system of present-day mammals, including *Homo sapiens*, has evolved
over millions of years. Because we do not know the nature of the sounds
that were important in earlier stages of the evolution of mammals, we do
not know the evolutionary pressures that were placed on different parts of
the developing auditory system. However, as we study the anatomy and
physiology of the ear and the auditory nervous system, we are surely examining a system that developed through many steps of evolution.

During the past two decades, our knowledge about the function of the
auditory system has increased rapidly through new techniques that have
come into use in neurophysiology and morphology. This book aims to
summarize the present state of our knowledge of the function of the ear
and the auditory nervous system. Emphasis is placed on presenting the
results of well-documented research in general, disregarding hypotheses
and theories not supported by experimental evidence. The first two
chapters are devoted to the physiology of the ear and auditory nervous
system; the third chapter describes the physiological basis for frequency
analysis in the auditory system. The fourth and final chapter discusses
coding of complex sounds in the auditory system. The book provides

more details on the function of the middle ear than what is found in available textbooks. In the treatment of the auditory nervous system, emphasis is placed on the analysis of complex sounds. Results obtained using simple test sounds are presented and compared with those obtained using complex sounds that resemble those of everyday life. This is different from other books on auditory physiology, which are usually restricted to experimental results obtained using very simple sounds. The present book also provides new ideas about the way the ear analyzes everyday sounds.

This book is intended for physiologists, audiologists, otologists, neurologists, and others who wish to be introduced to the field of auditory physiology. I hope that *Auditory Physiology* also will be useful for all advanced students in the fascinating field of hearing and communication.

Auditory Physiology

CHAPTER 1

Anatomy and General Physiology of the Ear

Introduction

This chapter is concerned with the general function of the ear and the auditory system. The auditory system is divided into the ear and the nervous system; the ear may be divided into the outer, middle, and inner ear as shown in the cross-sectional drawing of the human ear in Figure 1.1. The location of the ear in the skull is shown in Figure 1.2, whereas Figure 1.3 gives a schematic diagram of the ear as a whole. The outer and middle ears constitute the sound-conducting part of the ear, which transmits sound from air to the fluid of the inner ear. Thus, sound led through the external auditory canal sets the tympanic membrane into vibration, and these vibrations are transferred to the inner ear via the three small bones of the middle ear when the vibrations of the footplate of the stapes set the fluid in the cochlea into vibratory motion.

As discussed later in this chapter, the cochlea, and particularly the basilar membrane in the cochlea, plays an important role in analyzing the sound and converting it into a neural code. That code, after being modified in the different brain nuclei of the ascending auditory pathway, is transferred to the part of the cerebral cortex that receives auditory information. These transformations have not been completely studied, but our knowledge to date indicates that the information in the sound that reaches our ears undergoes substantial transformations. These matters are considered in Chapter 2.

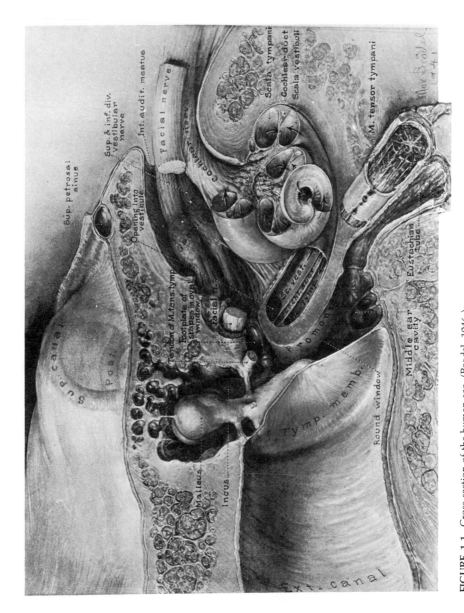

FIGURE 1.1. Cross-section of the human ear. (Brodel, 1946.)

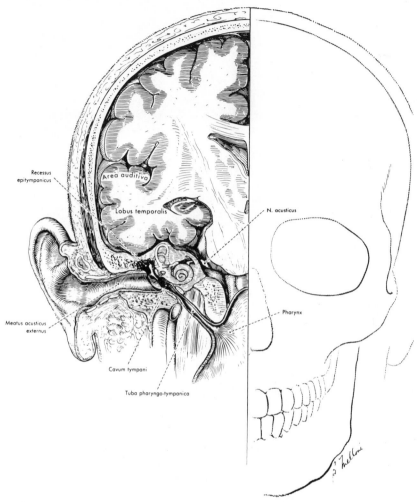

Recessus
epitympanicus

Area auditiva

Lobus temporalis

N. acusticus

Meatus acusticus
externus

Pharynx

Cavum tympani

Tuba pharyngo-tympanica

FIGURE 1.2. Location of the ear in the skull. (Based on Melloni, 1957.)

External Ear and Head

The external ear consists of the auricle and the ear canal. The external ear is shown in a schematic drawing in Figure 1.4. The groove called the *concha* is acoustically the most important part of the outer ear, whereas the flange that surrounds the concha is of little importance. Together with

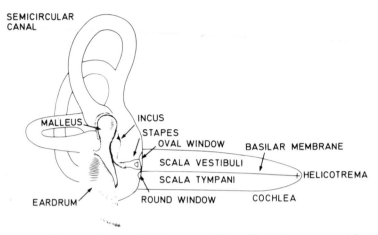

FIGURE 1.3. Schematic diagram of the human ear. The cochlea is shown as a straight tube.

the head, the outer ear and ear canal transform the sound field so that sound pressure at the tympanic membrane becomes different from the sound pressure as it is in the sound field where the head is situated (measured without the person being present).

All these components act as an integrated system that transforms sound from the free sound field to the tympanic membrane so that it is a function of both the frequency of the sound and the direction (azimuth) to the sound source.

PINNA

At higher frequencies, effects of the external ear are substantial acoustic gains so that sound pressure at the tympanic membrane is higher than it is in the sound field in which the head is situated. The effect of the pinna is mainly the result of resonances in the concha of the auricle.

EAR CANAL

The ear canal has a resonance around 3.5 kHz [i.e., the difference between sound pressure at the entrance of the ear canal and the tympanic membrane has its greatest value near that frequency (Figure 1.5)]. The resonance of the concha provides an acoustic gain at the entrance of the ear canal, relative to the sound field, of about 10 dB in the frequency range of 3–5kHz. Because the two resonators, the ear canal and the con-

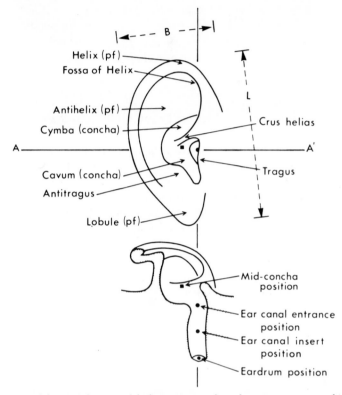

FIGURE 1.4. Schematic drawing of the human external ear showing components that are of importance for sound conduction: pinna flange (helix, antihelix, and lobule), concha (cymba and cavum), and ear canal. (From Shaw, 1974b.)

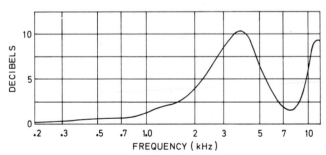

FIGURE 1.5. Effect of the ear canal on sound transmission. The difference in decibels between sound level at the entrance of the ear canal and that measured close to the tympanic membrane is shown as a function of frequency. (From Shaw, 1974a.)

cha, are coupled together, the total effect of the resonances is not the simple sum of the effect of the two studied in isolation. In addition to the
resonances just mentioned, the ear canal and particularly the concha
show resonances due to transverse models of vibration.

HEAD

The head acting as an obstacle in the sound field makes sound pressure
at its surface dependent on the angle to the sound source for frequencies
above 500–600 Hz. Acoustically, the head may be regarded as a hard surface (at least up to the upper limit of the audible frequency range), and a
valid model of the head is, in that respect, a sphere with a hard surface.
Such a sphere, equipped with two sound-pressure detectors or
microphones to simulate the ears, may be used to study the directionality
of the head (Figure 1.6). Two important observations can be made. One is
that the sound pressure at the surface of the sphere differs from that in a
free field only above a certain frequency, and that its effect is a function of
the *azimuth* angle between the sound source and the "head." When sound
directly impinges on one of the microphones, simulating the situation
where the sound source is located 90° azimuth, about a 6-dB higher sound
pressure can be measured for high frequencies compared to sound

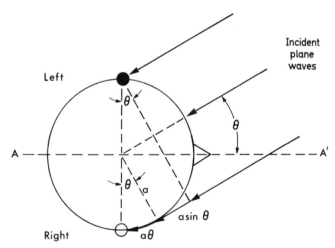

FIGURE 1.6. Schematic drawing of spherical model of the head with coordinate system for
horizontal plane through center of head showing azimuth of incident plane waves. (From
Shaw, 1974b.)

pressure in the free field (see Figure 1.7). This acoustic gain is primarily due to the baffle effect of the head. Another effect, the shadow effect, may be seen when observing the sound pressure on the opposite side of the head. There, a lower sound pressure than that in the free field, will be measured. This attentuation is present at high frequencies and for $-45°$ (and $-135°$) incidence, the sound pressure fluctuates with frequency between -3 and -6 dB (see Figure 1.7). For sound incidence at angles (azimuth) between these two extremes, the acoustic gain varies. When sounds come from a source located straight in front of or straight behind (grazing incidence) the head, their location of origin has very little effect

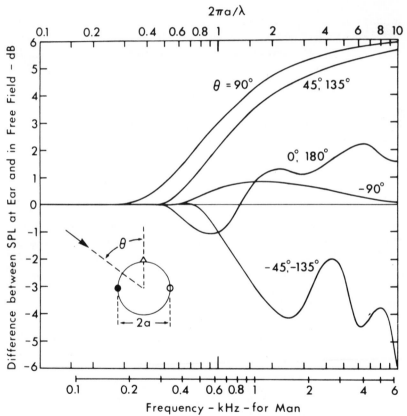

FIGURE 1.7. Calculated transformation of sound pressure level from free field to simple ear—a point receiver on the surface of a hard spherical model of the head as a function of frequency—for various values of azimuth (see Figure 1.6). (From Shaw, 1974b, based on Ballantyne, 1928.)

on the sound pressure and the acoustic gain fluctuates about the 0 dB axis.

These results favorably compare with results obtained in humans at least up to the 1.5–3 kHz region. The imperfect symmetry of the human head blurs the interference between the diffracted waves reaching the ear from the opposite side—the effects of which are the dips in the curve for −45° and −135° in Figure 1.7. The gain that results from the higher order resonances of the external ear is dependent on the angle of incidence of the sound, whereas the gain from the lower frequency resonances is practically independent of the angle of incidence of the sound.

The preceding discussion refers to a situation in which the sound source is located far away from the head. When the sound source, as in a normal listening situation, is moved closer to the head (source distance about 1–1.5 m), the gain at the ear that is nearer the sound source *increases*, and that at the remote ear *decreases*. At 1 m distance, the effect is about 1–2 dB (Hartley & Frey, 1921; Firestone, 1930).

COMBINED EFFECT OF HEAD, EAR CANAL, AND PINNA

Model studies and measurements of the sound pressure at different places in the human ear reveal that the resonance of the ear canal and the concha are particularly responsible for the acoustic gain from the sound in a free field to that at the tympanic membrane in the frequency range between 1.5 and 8kHz.

The total effect of the ear canal and concha resonance and that of the head for a sound source located in front of the person is shown in Figure 1.8 (Shaw, 1974a, b). This curve shows the acoustic gain relative to that in a free sound field. At around 2.5 kHz, this gain amounts to as much as 15 dB. Due to the effect of the angle of incidence shown in Figure 1.7, this gain function changes with the angle of incidence. (For further details about the acoustic properties of the external ear, see Shaw, 1974a, b.)

It is not only the magnitude of the sound pressure at the entrance of the ear canal that is a function of the direction to the sound source in a free sound field, but there is also an interaural time difference that is a function of the angle of incidence of the sound wave. That is a result of the fact that a sound wave from a distant source, except when the sound comes from a source located directly in front of the head or directly behind it, travels a different distance to reach the two ears. This difference in distance from the ears to the sound source is a function of the angle (azimuth) to the sound source (Figure 1.6), and thus the interaural time difference also

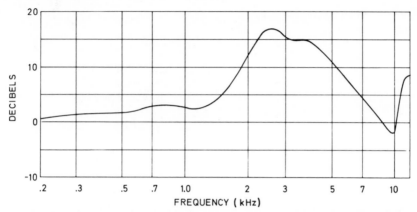

FIGURE 1.8. The combined effect of resonance in the ear canal and in the outer ear, and diffraction of the head. The curve shows the relative increase in sound pressure at the tympanic membrane with a person located in a free sound field with the sound source placed in front of the person, relative to the sound pressure at the same place but without the person being present. (From Shaw, 1974a.)

becomes a direct function of the azimuth. This is illustrated in Figure 1.9, where the solid line shows results obtained using a sphere with simple ears as described earlier. The circles show results obtained in a human subject. There is excellent agreement between the model results and the results obtained on a real head.

PHYSICAL BASIS FOR DIRECTIONAL HEARING

Dependence of the interaural time and intensity difference on the direction (azimuth) to the sound source is assumed to be the physical basis for directional hearing. Most work supports the hypothesis that it is the interaural time difference that is the most important factor for frequencies below 1500 Hz, whereas the intensity difference is the more important factor for frequencies about 1500 Hz. When discussing the physical basis for directional hearing, two additional factors are of importance—namely, the fact that natural sounds usually have a broad spectrum and that the head is usually turned during the process of localizing a sound source. The latter makes it possible to use the change in interaural time difference and the change in interaural intensity difference instead of relying on absolute values of these differences.

Localization of a sound source in the vertical plane for sound sources located at different elevations is assumed to depend on differences in the

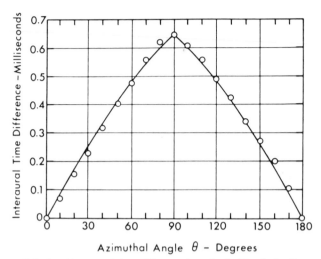

Azimuthal Angle θ – Degrees

FIGURE 1.9. Calculated interaural time difference for spherical head of radius a = 8.75 cm (solid line) as a function of azimuth (incident plane of sound wave) compared with average measured values (open circles) for a human subject (see Figure 1.6). (From Shaw, 1974b, after Feddersen *et al.*, 1957.)

transformation of sound intensity from the free sound field to the eardrum. Although not studied in detail, it is generally assumed that the pinna plays an important role in the localization of a source in the vertical plane by modifying the spectrum of a broadband sound as a function of its elevation.

When the ear is artificially stimulated, such as by an earphone, the physical basis for directional hearing is naturally lost because the head does not produce an interaural time and intensity difference. In addition to that, and sometimes more importantly, the earphone modifies the resonance effect of the outer ear (canal and concha). An earphone, such as those used in hearing aids, inserted in the ear eliminates the effect of the concha and modifies the effect of the ear canal. The result is a loss of the acoustic gain at high frequencies normally obtained by the external ear. An earphone with a cushion, such as those used in clinical audiometry and stereo listening, compresses the concha and modifies the ear canal resonance.

Knowledge about the acoustic effect of the external ear is also important when considering the damaging effect of industrial noise on the ear. In estimating the risk for such damage, the noise level is usually measured in a free field using noise level meters that do not take into account the frequency-dependent acoustic gain of the head and external ear.

Middle Ear

ANATOMY

Figure 1.10 presents a schematic representation of the middle ear. The tympanic membrane is a cone-shaped membrane connected to the manubrium of the *malleus* over a length of a few millimeters. The malleus in turn is connected to the *incus.* The short process of the incus rests in a bony fossa in the posterior floor of the epitympanic recess, whereas its long process is connected to the stapes through the incudostapedial joint. The joint is most flexible for movements perpendicular to the pistonlike movement of the stapes. The *footplate* of the stapes rests in the oval window of the inner ear (cochlea). The tympanic membrane is under slight inward tension and its radial and circular fibers effectively stiffen it while contributing little additional mass. The malleus is suspended by several

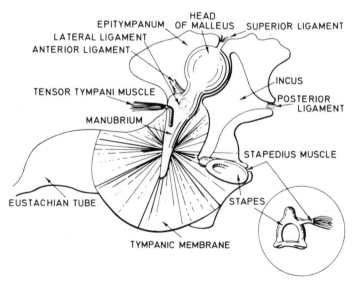

FIGURE 1.10. Schematic drawing of the middle ear of a human seen from inside the head. (From Møller, 1972.)

ligaments. In man, it is usually in firm contact with the incus, although in many animals this joint allows some movement.

There are marked differences between the physical dimensions of the human middle ear system and those of the commonly used experimental animals such as guinea pigs, cats, rabbits, rats, and monkeys (Fumagalli, 1949), but the movements of the ossicles seem essentially similar.

The muscles of the middle ear are discussed in the fourth section of this chapter. The middle ear is a closed cavity except for the Eustachian tube leading to the nasopharynx. This tube opens during swallowing, thus ventilating the middle ear.

GENERAL FUNCTION OF THE MIDDLE EAR

In order to reach the sensory cells in the cochlea—a part of the inner ear—sound must be transformed into vibrations in the fluid of the cochlea. Direct transmission of sound from the air to the fluid in the inner ear is extremely inefficient because of the great difference in mobility of the two media. This results in a large reflection when sound impinges upon such a fluid surface. The fluid in the inner ear can be regarded as having properties similar to those of water, and when sound reaches such a water surface directly as much as 99.9% of the energy is reflected. The middle ear enhances the transformation of sound into vibrations of the cochlear fluid by acting as a mechano-acoustic transformer that equalizes the high impedance of the inner ear and the low impedance of air. The impedance of air is 42 cgs units and that of water is 1.54×10^5, giving a ratio of about 1 : 4000. Thus, the transmission of sound to the oval window will be optimal if the middle ear has a transformer ratio of 1 : 4000. In such a case, the impedance of air will perfectly match that of the cochlea (seen from the oval window), and the transmission of sound will be improved by 30 dB over that when sound is transmitted directly to the oval window. This value of 30 dB is the theoretical maximal value, and it is not realized in practice, however, for several reasons.

The *transformer ratio* of the human middle ear is slightly different from that required for optimal transfer of sound to the inner ear. The transformer action of the middle ear is mainly accomplished by the ratio between the areas of the tympanic membrane and the stapes and to a lesser degree by the lever ratio of the ossicles. In addition, it is somewhat different for different frequencies, mainly because the functional area of the tympanic membrane is frequency dependent (Wever & Lawrence, 1954).

The elasticity and mass of the middle ear introduce stiffness an inertia, respectively, into the system. Both these components resist motion to a degree that is frequency dependent. The effect of inertia increases with frequency, whereas that of stiffness is greatest at low frequencies.

Even so, the middle ear can transfer sound to the inner ear relatively efficiently over the entire audible range (i.e., from about 10 to 20,000 Hz in man). The range extends much higher in many mammals (e.g., up to about 50 kHz both in the cat and the rat, and to over 100 kHz in the bat).

In evaluating the functional gain of the middle ear, it should also be taken into consideration that it is the *difference* in force at the two windows that sets the cochlear fluid into motion. The transformer action of the middle ear renders the direct force of the oval window much greater than that on the round window. That is of great importance in setting up an efficient difference in force at the two windows. If sound reached both windows with the same intensity, as it likely would if the middle ear were absent, no movement of the cochlear fluid would result. Loss of the middle ear function would also result in sound reaching both the oval and round windows with about the same intensity and phase angle. Therefore, loss of the middle ear function as a result of disease may result in a hearing loss of 50–60 dB, an amount much greater than the gain of the middle ear in transmitting sound to the oval window.

The capacity for the middle ear to improve the transmission of sound from the air to vibration in the cochlear fluid has been studied experimentally in animals (see Wever & Lawrence, 1954) and in human cadaver ears (von Békésy, 1941).[1] Figure 1.11 shows the sound pressure normally required at the tympanic membrane of a cat to produce a certain small movement in the cochlear fluid (measured by recording the cochlear microphonic potential; see the section on electrical potentials in the cochlea p. 89), compared to the sound pressure required when the middle ear has been removed. The sound was led to either the oval or the round window. The graph in Figure 1.11 shows that a greater sound pressure is required to produce the same degree of fluid motion under the latter circumstances. Also, this graph shows that sound transmission to either of the cochlear windows is equally efficient in setting the cochlear fluid into motion. The difference between the curves representing the efficiency of the middle ear in enhancing the transmission from air to the

---

[1] The work of von Békésy is collected in the form of a book (*Experiments in Hearing,* McGraw-Hill, 1960), but the various works will be referenced here as the original journal papers.

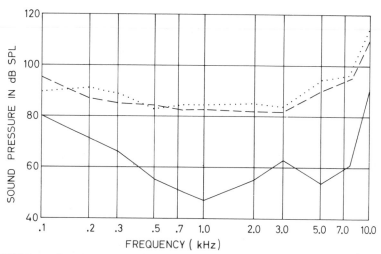

FIGURE 1.11. Sound pressure necessary to produce a certain cochlear microphonic potential (10μV) when the sound is delivered to the tympanic membrane (solid line) with the middle ear intact and when it is delivered to the round window (dots) or the oval window (dashes) with the middle ear removed. The cochlear microphonic potential can be recorded at the round window, and equal amplitude of this potential is assumed to be an indication of equal vibration velocity of the cochlear fluid. The results are from an experiment in a cat. (From Wever, E.G., Lawrence, M., & Smith, K.R. The middle ear in sound conduction. *Archives of Otolaryngology,* 1948, *48,* 19–35. Copyright 1948, American Medical Association.)

cochlear fluid is shown in Figure 1.12 as a function of frequency. The middle ear is more efficient in the middle frequency range than at relatively low or high frequencies.

Properties of the human middle ear have also been measured in

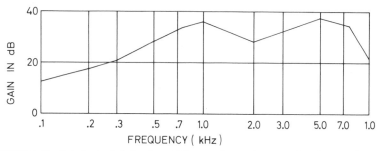

FIGURE 1.12. Gain in decibels by the middle ear of a cat determined as the difference between the sound necessary to elicit a certain amplitude (10μV) of cochlear microphonic potential when led to the round or oval window compared with leading the sound to the inner ear through the middle ear. The graph shows the difference between the curves in Figure 1.11.

cadaver ears. Although the changes that occur after death are not known in detail, it may be concluded that the middle ear in man enhances the transmission of energy to the oval window by 25 dB at the most. This optimum transmission occurs in the frequency range of 500–5000 Hz.

TRANSMISSION PROPERTIES

Frequency dependent transmission properties of the middle ear are usually expressed as a transfer function.[2] The transfer function of the middle ear can be either the ratio between the vibration amplitude of the stapes and the sound pressure at the tympanic membrane as a function of frequency, or the ratio between the vibration velocity of the stapes and the sound pressure at the tympanic membrane. Figure 1.13 shows the transfer function of a cat's middle ear as observed in an investigation using microscopic measurements of vibration amplitude of the stapes of an anesthetized cat (Guinan & Peake, 1967). This graph shows the vibration amplitude of stapes for constant sound pressure level at the tympanic membrane. These results are in agreement with the results of other studies by Møller (1963), in which the measurements were made using a capacitive probe. These animal experiments thus indicate that the vibration amplitude of the stapes for constant sound pressure level at the tympanic membrane is independent of frequency up to a certain point, above which it falls with a certain slope. It is generally agreed that the first part of this slope approximates 18 dB/octave. Above a certain frequency, however, the course of the transfer function becomes less regular and probably less steep, as well. Thus, in transforming sound pressure into the vibration amplitude of the stapes, the middle ear generally acts as a lowpass filter.

The transfer function shown in Figure 1.13 shows the vibration amplitude for constant sound pressure at the tympanic membrane. Another measure, such as velocity of the stapes may be more appropriate to use when assessing the transmission properties of the middle ear. Volume velocity of the cochlear fluid may be more closely related to the excitation in the cochlea than to the volume displacement. Transfer func-

[2] The properties of a transmission system like the middle ear are often expressed by its *transfer function*. That consists of a *gain function*, which is the ratio between the output and the input when the input is sinusoidal, and a *phase function*, which is the phase shift between the output and the input. Both the gain functions and phase functions are usually plotted as a function of frequency, and when the gain function is expressed in logarithmic values, usually decibels, it is called a *Bode plot*. The transfer function can also be viewed as a curve that shows how the output varies as a function of frequency when the input amplitude is kept at a constant level. In a linear system the transfer functions are identical for different input amplitudes to the system.

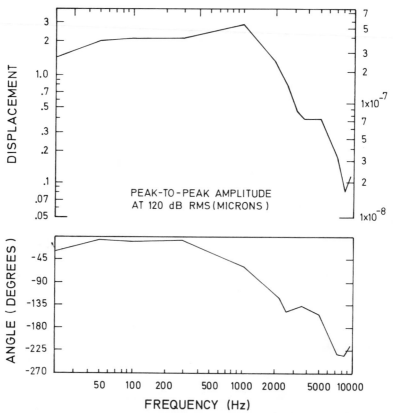

FIGURE 1.13. Displacement of stapes in an anesthetized cat for a constant sound pressure level at the tympanic membrane (120 dB SPL). The sounds were pure tones at different frequencies. Both the amplitude of the displacement and the phase angle between the sound and the displacement are shown. The vertical scale of the upper graph is in micrometers peak-to-peak for 120 dB SPL at the tympanic membrane. (From Guinan & Peake, 1967.)

tion, with regard to displacement for a sinusoidal motion, can be converted to a transfer function for velocity by tilting the gain function with 6 dB per octave. (The velocity is the derivative of the displacement.)

It is important that the transfer function of the middle ear be determined with the middle ear intact, including the cochlea. The cochlea is important since the cochlear fluid acts as a load on the stapes footplate. The cochlear fluid is thus a significant part of the mechanical system of the middle ear. In some experiments, the vibration of the stapes has been measured with the cochlear fluid removed. Results of such experiments do not give a correct picture of the properties of the middle ear in transforming sound into vibration of the cochlear fluid.

IMPULSE RESPONSE FUNCTION OF THE MIDDLE EAR

Impulse response of a system is its response to a very brief impulse. Direct measurements of the impulse response of the middle ear were made by von Békésy (1936) in human cadaver ears. He recorded the motion of the malleus in response to a brief impulse. Direct determination of the impulse response function of the middle ear is, however, technically difficult, and it has therefore been done using an indirect method based on the transfer function measured using pure tones. Figure 1.14 shows the impulse response of the middle ear of a cat, computed from the transfer function from the sound pressure at the tympanic membrane and vibration amplitude of the round window measured using a capacitive probe (Møller, 1963; see also Figure 1.13). This computation of the impulse response may be taken as a reliable measure of the displacement of the

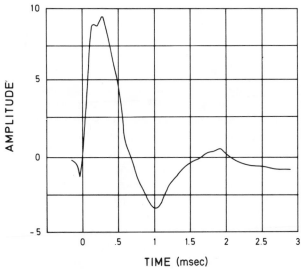

FIGURE 1.14. Impulse response function of the middle ear of a cat. The curve is calculated vibration amplitude of the round window in response to a short sound impulse delivered to the tympanic membrane. The calculation consisted of taking the inverse Fourier transform of the frequency domain transfer function obtained using sinusoidal stimulation. (From Møller, 1972.)

cochlear fluid. Since the middle ear can be regarded as a linear system over the physiological range of sound intensities, it is appropriate to compute its impulse response on the basis of a frequency domain transfer function. The impulse response shown in Figure 1.14 can thus be regarded as the displacement of the round window in response to a brief impulsive sound at the tympanic membrane.

DETAILED DESCRIPTION OF THE ACOUSTIC PROPERTIES

The middle ear transmits different frequencies differently because of the variation in the mass and the elasticity of the components of the middle ear that participate in vibrations. A third quantity—friction—is also important to consider. The relative influence of each of these three components on the function of the entire system can be illustrated by considering their *mechanical impedances*. When a sinusoidal force is applied to a mechanical system, its motion is a forced oscillation. The amplitude, or rather velocity of the resulting motion is a function of the force and the impedance of the system. More specifically, the mechanical impedance is defined as the ratio between the force and the resulting velocity. Mechanical impedance of a mass is a result of its inertia and increases proportionally with frequency, whereas the impedance of an elastic component decreases with frequency. However, the impedance of these two components is a reactive or imaginary impedance, and it has opposite signs for inertia and stiffness. A simple system consisting of only one mass and one stiffness (Figure 1.15A) possesses, then, a certain frequency at which impedances of these two quantities are numerically the same. At that frequency, which is called the system's resonance frequency, impedance of the inertia and the stiffness, being opposite in sign, cancel one another. The system thus theoretically attains no impedance at that frequency. Most systems include a third component—friction—the impedance of which is independent of the frequency. Its value cannot be added directly to the reactive impedance of inertia and stiffness, but must be combined with the impedance of the mass and resistance as a vector shifted 90° to obtain the impedance of the total system (see Figure 1.15B).[3] In a system containing friction, there is not a complete cancellation of the impedance at the resonance frequency but at the resonance frequency the

[3] The energy demanded to set a mass into motion or to deform a spring is stored in the system as either kinetic or positional energy, and theoretically can be recovered at any time. Friction and resistance, on the other hand, convert the mechanical or acoustic energy into heat.

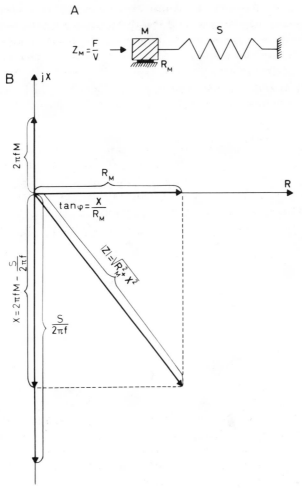

FIGURE 1.15. (A) Simple mechanical system consisting of a mass (M), elasticity (S), and friction (R_M); (B) relationship between the different elements of the impedance ($Z = R + jX$) and frequency (f) of the system shown in Part A of this figure. (From Møller, 1964a.)

impedance of the system is that of the friction component. Consequently, the minimal impedance of a system containing friction occurs at the resonance frequency, but its sharpness of the resonance is proportionate to the magnitude of the friction component.

It should be emphasized that the mechanical system shown in Figure

1.15A is a very simple system. Most systems are far more complicated, and contain many more elements. The mechanical system in Figure 1.15A can only be regarded as a very simplified model of the inertia and friction of the middle ear. The inverse of impedance is called admittance, and it is a measure of the mobility of the system.

In order to understand the function of the middle ear, it is practical to extend the simple mechanical model shown in Figure 1.15A. The mechanical system in Figure 1.15A was transformed into an acoustico-mechanical system (see Figure 1.16). In this latter system, sound waves reaching the piston apply a force to the mass in a way, similar to that in which a mechanical force shown in Figures 1.15A and 1.15B would. The impedance of the system in Figure 1.15A is now an acoustic impedance, and it is defined as the sound pressure divided by volume velocity of the piston. In the ideal system shown in Figure 1.16A, the acoustic impedance is the mechanical impedance divided by the square of the area of the piston. That assumes an ideal model in which the piston has no mass and no compliance.

It is common to illustrate mechanical or acoustic systems by their electrical analogs. Such models often become more illustrative than mechanical models, and people are more familiar with electrical circuits than with acoustic or mechanical systems. Figure 1.16B shows an elec-

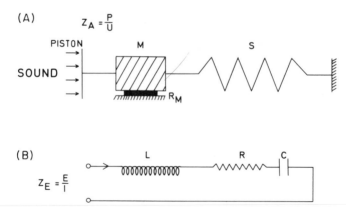

FIGURE 1.16. The mechanical system seen in Figure 1.15A equipped with a rigid piston to form an acoustical system (A) together with its electrical analog (B). Z_A is the acoustic impedance and Z_E is the electrical impedance. P is sound pressure, U is volume velocity, E is voltage, and I is current. The inductance L represents the inertia of the mass M, the capacitance C represents the stiffness of the spring S, and the resistance R represents the friction component R_M. (From Møller, 1964a).

trical model of the mechanical system in Figure 1.16A. In an electrical model, inertia (M) is represented by an inductance (L), friction (R_M) by a resistance (R), and elasticity (S) by capacitance (C). Force applied to a mechanical system corresponds to voltage applied to the electrical analog circuit. An electrical current has velocity as its counterpart in the mechanical system. Consequently, mechanical impedance, for example, force divided by velocity, corresponds to electrical impedance, voltage divided by current. (The reader who is interested in acoustic and electrical analogies is referred to Olson, 1958.)

The middle ear is more complicated than the acoustico-mechanical systems illustrated in Figure 1.16A. The simplified model of the middle ear shown in Figure 1.17 contains both series and parallel components. One of the shunt elements (C_4, R_4) in the electrical analog circuit is a result of elasticity in the incudostapedial joint, and another shunt element (R_2, C_2) is caused by the impedance of the tympanic membrane.

An important implication of shunt elements follows directly from an examination of the electrical analog. In a system without shunt elements, a simple relationship exists between the transmission properties and the impedance. *Transmission* is defined as the ratio between the current that flows in the resistor in the model in Figure 1.16B and the input voltage. In a simple system as shown in Figure 1.16B, the output current is a direct function of the input voltage and the impedance of the system. With these two quantities known, the transmission can be computed. In the system in Figure 1.17 this is not the case.

If the model of a simple mechanical system in Figure 1.16 was valid for the middle ear, the velocity of the cochlear fluid would be the sound

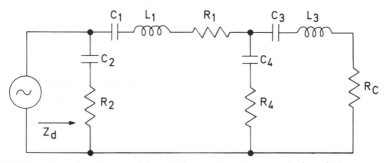

FIGURE 1.17. Analog diagram of the middle ear. As in Figure 1.16B, the inductances represent inertia, capacitances stiffness, and resistance represents friction. (From Zwislocki, 1963.)

pressure at the tympanic membrane, divided by the impedance of the middle ear measured at the tympanic membrane, and a constant factor depending on the ratio of the area of the tympanic membrane to that of the stapes footplate. In circuits where there are shunt elements, the situation is more complicated, and the transmission properties and the impedance of the middle ear bear no simple relationship to each other. This is shown by the electrical analog (Figure 1.17). In this case some of the current is lost in these shunt elements, so that its transmission properties cannot be exactly predicted on the basis of its impedance. This will be considered in detail later in this chapter.

The Acoustic Impedance of the Ear

Acoustic impedance at the tympanic membrane is a measure of the "resistance" that the tympanic membrane exerts against being set into motion by a sound wave. It represents the inverse mobility of the tympanic membrane, the middle ear bones, and the cochlear fluid as seen from the stapes. If the tympanic membrane were to function as a rigid piston, the acoustic impedance would be the mechanical impedance (of the manubrium of the malleus) divided by the squared area of the piston as illustrated in Figure 1.16A.

Measurements of the acoustic impedance of the ear can be performed in intact ears and have played an important role in studies of the function of the middle ear (Zwislocki, 1957; Møller, 1961a, 1963, 1964a, 1965). Figure 1.18 shows a comparison betwen the impedance of a typical human ear, a cat ear, and a rabbit ear. It is shown that certain differences exist between the impedances of these commonly used experimental animals and those of man.

Impedance curves plotted as a function of frequency in the human ear have more peaks than impedance curves of the ear of cats or rabbits. Compared to these animals, this indicates a more complex vibration pattern and possibly a less efficient action in the human tympanic membrane. Large individual variation exists in the impedance of the ears of different individuals (Figure 1.19), even though the impedances may be all obtained in young subjects with normal hearing. Individual variation has greatly hampered the use of acoustic impedance as a tool to diagnose middle ear diseases, but it remains a valuable tool in animal experiments on the function of the middle ear.

In order to facilitate the interpretation of acoustic impedance data in

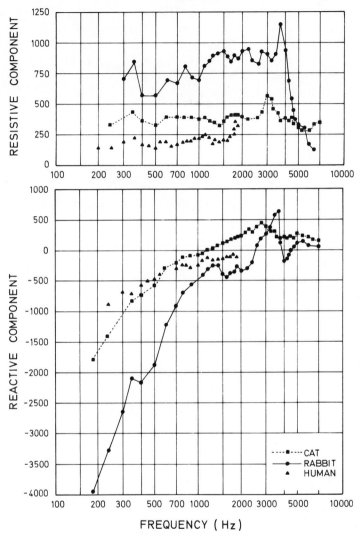

FIGURE 1.18. Acoustic impedance at the tympanic membrane of a typical human (triangles), cat (squares), and rabbit (circles) ear. The resistive part is shown in the upper graph, and the reactive part is shown in the lower graph. (From Møller, 1972.)

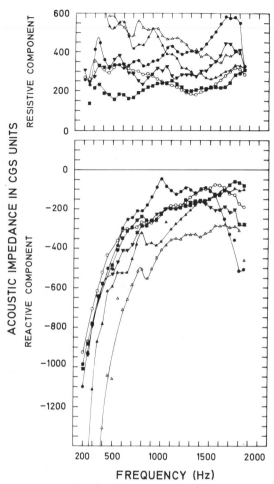

FIGURE 1.19. Acoustic impedance of the normal ears of six individuals with normal hearing. (From Møller, 1961a.)

terms of transmission properties, it is valuable to consider the effects of the various parts of the middle ear. As mentioned earlier, it is only when the effect of the shunt elements in the model ear can be neglected that the middle ear's transmission properties can be directly deduced from the acoustic impedance measured at the tympanic membrane. When admittance of the shunt elements is of significant magnitude, their effect must be considered if the transmission properties are to be determined on the basis of measured acoustic impedance of the tympanic membrane.

FUNCTION OF THE TYMPANIC MEMBRANE

In converting sound into mechanical vibration, the efficiency of the tympanic membrane is a function of the degree to which its vibrations are transferred to the manubrium of the malleus. Measurements of acoustic impedance show that the cat's tympanic membrane functions much in the same way as a rigid piston for frequencies up to about 3 kHz (Møller, 1963, 1965).

Comparison of the acoustic impedance with the mechanical impedance at the malleus is one way of studying the function of the tympanic membrane. Since the mechanical impedance is the force divided by the resulting velocity, measurement of the velocity of the malleus at a constantly applied force gives the mechanical impedance. When the tympanic membrane functions as a piston, the effects of a constant sound pressure level at the tympanic membrane are identical to those of applying a constant force to the malleus.

Results obtained by comparing the acoustic impedance measured at the tympanic membrane and the mechanical vibratory properties of the middle ear of a cat are shown in Figure 1.20. Acoustic impedance at the tympanic membrane is shown together with the ratio between the sound pressure at the tympanic membrane and the vibration velocity of the malleus are shown in Figure 1.20 as a function of frequency. Both amplitude data (in decibels) and phase data (in degrees) are shown. Results of measuring the vibration amplitude of the malleus are expressed as relative values adjusted so that the two curves fit at low frequencies. Such results provide important information about the frequency dependence of the effective area of the tympanic membrane and are likely to mean that the efficient area of the tympanic membrane is relatively constant for low frequencies, and that the shunt elements do not influence sound transmission for low frequencies. The difference in the two functions above 3 kHz (Figure 1.20) indicates that the efficient area of the tympanic membrane is different at these frequencies and that there may be an effect of elastic elements that function as shunt elements in accordance with the models shown in Figure 1.17.

It is crucial to consider the influences of the tympanic membrane when the function of the middle ear is being investigated through acoustic im-

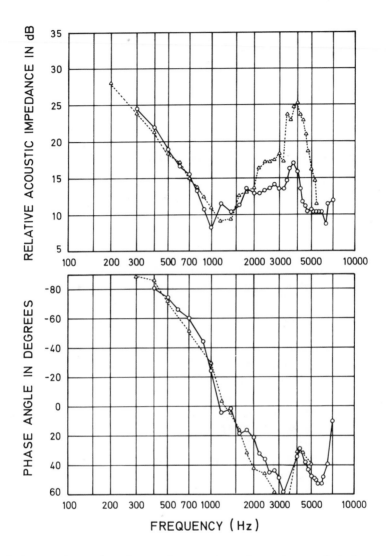

FIGURE 1.20. Acoustic impedance at the tympanic membrane of an anesthetized cat compared with the inverse vibration velocity of the malleus at constant sound pressure at the tympanic membrane. The latter is the mechanical impedance of the middle ear provided that a constant sound pressure is equivalent to applying a force of constant amplitude to the malleus. The upper graph shows absolute values in decibels, and the lower graph shows the phase angle in degrees. The bulla was open and the bony septum in the middle ear was removed. The reference level of the acoustic impedance was 100 cgs units, whereas it was arbitrary for the inverse velocity. (From Møller, 1963.)

pedance measurements. What is measured as acoustic impedance is actually the mechanical impedance of the middle ear, transformed into acoustic impedance by the tympanic membrane and the impedance of the tympanic membrane itself. The ideal situation, therefore, occurs when the tympanic membrane functions as a rigid piston, rendering the mechanical impedance equal to the acoustic impedance multiplied by a constant frequency-independent factor.

Impedance of the tympanic membrane may be studied by comparing the acoustic impedance before and after the malleus has been immobilized. Results of such experiments measure the mobility of the tympanic membrane itself. The results of such an experiment in a cat are shown in Figure 1.21. Impedance of the tympanic membrane itself, measured when the malleus is immobilized, is high for frequencies up to about 2500 Hz. The impedance of the tympanic membrane itself above 3 kHz is about the same as that of the intact ear, indicating that large parts of the tympanic membrane vibrate without the vibration being transferred to the manubrium of the malleus. These results indicate that the tympanic membrane functions with optimal efficiency up to 2 kHz. Above that frequency, increasingly larger parts of the tympanic membrane vibrate without the vibrations being transferred to the malleus. Similar, reliable data on the human ear are difficult to obtain since only cadaver ears are available for such studies and animal experiments have shown that the acoustic properties of the middle ear change after death. However, acoustic impedance data indicate that the human tympanic membrane is less stiff than the cat's and may therefore be less efficient.

Von Békésy (1941) measured the vibratory amplitude of the different parts of the tympanic membrane in human cadaver ears at the frequency of 2400 Hz. Measurements have been repeated and improved by Tonndorf and Khanna (1970) using laser holography (Figure 1.22). Figure 1.22 leaves no doubt as to the complexity of the vibration of the tympanic membrane.

The conical shape of the tympanic membrane makes it stiffer than a flat membrane of the same weight. The conical shape also allows the tympanic membrane to vibrate as a unit with less decrease than would be the case for a flat membrane in vibration amplitude toward the edges. It was earlier thought that the conical shape of the tympanic membrane gave rise to nonlinearities. Newer results from animal experiments indicate that this is not the case for physiological sound levels (Guinan & Peake, 1967).

Since it is the *difference* between the pressure on the two sides of the tympanic membrane that causes it to move, it is important that as little of

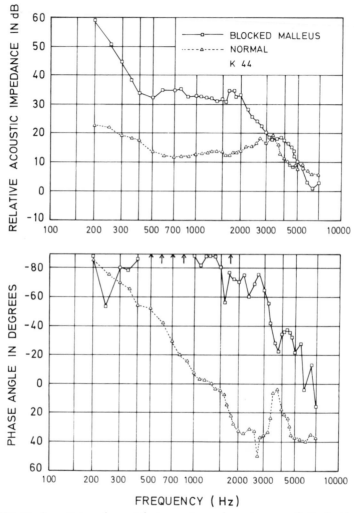

FIGURE 1.21. Acoustic impedance at the tympanic membrane in an anesthetized cat before (triangles) and after (squares) the ossicular chain has been immobilized by gluing the malleus to the wall of the middle ear. The bulla was open and the bony septum in the middle ear was removed. The upper graph shows the absolute value of the impedance (in decibels re 100 cgs units), and the lower graph shows the phase angle of the impedance. (From Møller, 1965.)

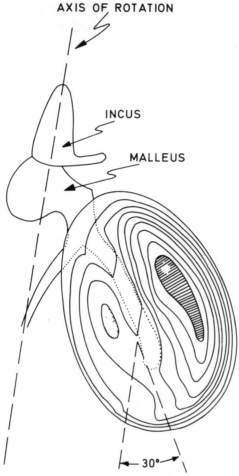

AXIS OF ROTATION

INCUS

MALLEUS

30°

FIGURE 1.22. Vibration of the tympanic membrane determined using holography. (From Tonndorf & Khanna, 1970.)

the sound as possible reaches the middle ear cavity. This is normally the case, but if the tympanic membrane becomes perforated, as it may in certain pathological conditions, sound will reach the middle ear cavity through this hole and hearing will then be impaired because the resultant force available to move the tympanic membrane decreases. A perforation acts as an inertial component that transmits sound of low frequency to the middle ear cavity, whereby the efficient force on the tympanic membrane decreases for these frequencies. Small holes only transmit very low frequencies, but increasing the size of the hole will allow it to transmit sound of higher frequencies. A small hole will cause impaired hearing only at very low frequencies, whereas a larger hole will cause hearing loss over an increasingly larger frequency range (Tonndorf, McCardle, & Kruger, 1976). If large parts of the tympanic membrane are missing, its function as a piston will also deteriorate and the hearing loss in this case will increase both in magnitude and the range of frequencies at which it extends.

IMPEDANCE OF THE COCHLEA REFLECTED AT THE TYMPANIC MEMBRANE

Studies in the cat and rabbit indicate that the cochlea contributes most of the frictional component of the impedance that can be measured at the tympanic membrane, whereas very little of the reactive component of the impedance seems to originate from the cochlea (Møller, 1965). These results were obtained by measuring the ear's acoustic impedance before and after the incudostapedial joint was interrupted, thus disconnecting the cochlea from the middle ear. Examples of the results of these experiments are shown in Figure 1.23. Shown is the acoustic impedance measured at the tympanic membrane of a cat when the middle ear was intact compared to when the incudostapedial joint was interrupted. A more pronounced dip in the impedance is seen when the inner ear is disconnected, indicating that the resonance is sharper in the absence of the inner ear than when the inner and middle ears are connected. The mechanical system of the middle ear is thus much less damped, since the friction in the system is reduced when the influence of the cochlea is eliminated. This is clearly seen when the impedance is expressed in its reactive and resistive components (Figure 1.24). Over a large range of frequencies, the resistive component falls to nearly zero, whereas the reactive component changes relatively little as a result of disconnecting the cochlea.

Results of experimentation thus show that the middle ear has very little friction and that the cochlea "seen" from the oval window is almost a pure friction. The allocation of the greater part of the frictional component of the ear's impedance to the cochlea is a prerequisite for the optimal transfer of acoustic energy to the cochlea. The conclusion that the impedance of the cochlea is likely to be nearly a pure frictional resistance, as seen from the oval window, has also been drawn from theoretical analysis of the function of the cochlea by Zwislocki (1948).

Direct measurements of the impedance of the cochlea entail great technical difficulties, and the results of such studies vary widely.

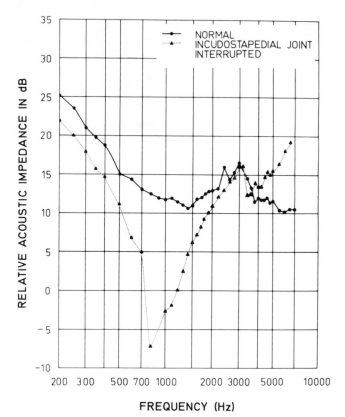

FIGURE 1.23. Effect of interrupting the incudostapedial joint on the acoustic impedance at the tympanic membrane in a cat. Acoustic impedance is given in decibels (re 100 cgs units). (From Møller, 1965.)

However, these results generally confirm that the cochlea can be regarded as a source of almost pure resistance over most of the audible frequency range (Tonndorf, Khanna, & Fingerhood, 1966).

SOUND TRANSMISSION THROUGH THE INCUDOSTAPEDIAL JOINT

The incudostapedial joint has a large amount of elasticity, and it can therefore be expected to be inefficient in transmitting sound (vibration) at high frequencies. In the electrical analog of this joint (Figure 1.17), elasticity corresponds to shunt capacitance. The effect of this elasticty has been studied in animal experiments by comparing the vibration velocity

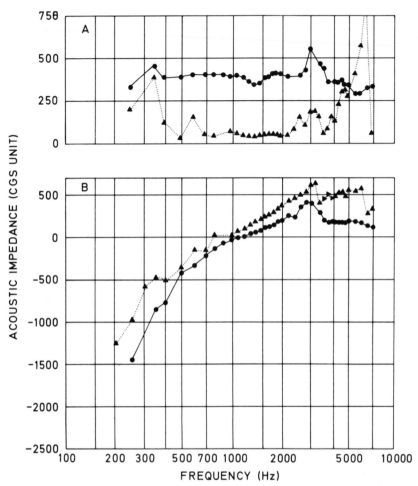

FIGURE 1.24. The same data as shown in Figure 1.23 but with resistive (A) and reactive (B) components of the impedance shown separately. (From Møller, 1965.)

of the incus and the round window (Møller, 1963). Such measurements show a discrepancy in amplitude above 2500 Hz which increases to about 5 dB at 5 kHz. Figure 1.25 shows the relative vibration amplitude of the incus compared with that of the round window. The latter is assumed to represent the vibration of the stapes, since the cochlear fluid can be regarded as incompressible and the cochlear capsule can be regarded as a rigid wall. The discrepancy between these two functions is assumed to result from the effect of the elasticity in the incudostapedial joint.

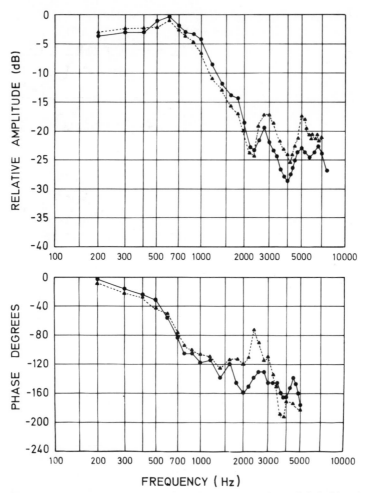

FIGURE 1.25. Relative vibration amplitude of the incus (triangles and dashed lines) compared with that of the round window (circles and solid lines) in an anesthetized cat. Results were obtained using a capacitive probe, and the two curves were shifted vertically to the best fit for low frequencies. (From Møller, 1963.)

One more interesting result that emerges from Figure 1.25 is that the vibration amplitude does not continue to decrease with increasing frequency above its resonance frequency as it would be expected to do if the middle ear could function as a simple combination of mass, elasticity, and friction. Instead, the vibration amplitude increases above 2500 Hz and then fluctuates around a certain value for frequencies above that of the principle resonance frequency of the middle ear. This behavior shows better transmission than predicted by the current models of the middle ear. The reason for this is unknown, but it is also seen in other studies (see Guinan & Peake, 1967).

EFFECT OF THE MIDDLE EAR CAVITIES

In general, the middle ear cavity acts as an air cushion behind the tympanic membrane, impeding its motion. The exact way that this occurs depends on the frequency of the sound as well as on the anatomy of the middle ear cavity, which varies among animal species. In animals that have a single cavity, such as the rabbit, it adds stiffness to the middle ear. However, the complicated dual cavity of the cat's middle ear leads to resonance. Because the middle ear cavity of the cat is divided by a *septum* with a small hole, it vibrates mechanically like a Helmholtz resonator (Figure 1.26). This resonator is sharply tuned, and its effect, therefore, is small outside the narrow frequency range around its resonance frequency. At the resonance frequency (about 4 kHz), the cavities exert a high resistance to movement of the tympanic membrane. Near that frequency the acoustic impedance at the tympanic membrane undergoes a sharp increase shown in Figure 1.27. The transmission of sound to the cochlea in the cat is therefore reduced in a narrow frequency range around the resonance frequency of the middle ear cavities. Because the resonance frequency closely coincides with one of the common audiometric frequencies (4 kHz), behavioral threshold values in the cat are rendered extremely variable at that frequency.

In man, the middle ear cavity is relatively large and often a number of air cells communicate with the main cavity. Little is known about the exact acoustic function of these air cells, but they probably add some friction to the middle ear's acoustic system, influencing its transmission properties at higher frequencies. However, because the middle ear cavity is relatively large, air cells at low frequencies are likely to produce only a small effect on the transmission properties of the middle ear.

EFFECT OF AIR PRESSURE

It is acknowledged that the inequality of air pressure between the two sides of the tympanic membrane causes elevation of the hearing threshold in man. Normally, air pressure in the middle ear cavity is kept close to that of the environment through the action of the Eustachian tube that

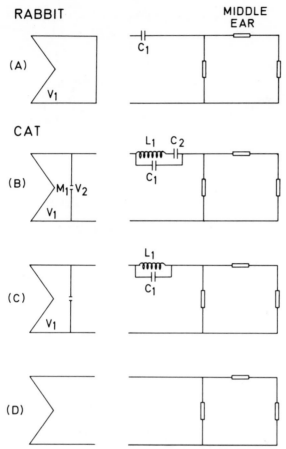

FIGURE 1.26. Schematic illustration of the middle ear cavities in the rabbit and the cat, together with their electrical analogs: (A) the situation in a rabbit with the bulla intact; (B) the cavity system in the cat with the bulla intact; (C) the cat with the outer cavity opened; (D) the cat with the bony septum removed. The right-hand network on each drawing represents the respective electrical analog. (From Møller, 1965.)

opens briefly when one swallows or yawns. A slight difference in air pressure elevates the threshold at low frequencies more than at high frequencies, but a large difference in air pressure can affect the threshold even at high frequencies (Rasmussen, 1948). The effect of air pressure is believed to result from an increase in stiffness in the middle ear, probably due to distention of the tympanic membrane. If the tympanic membrane

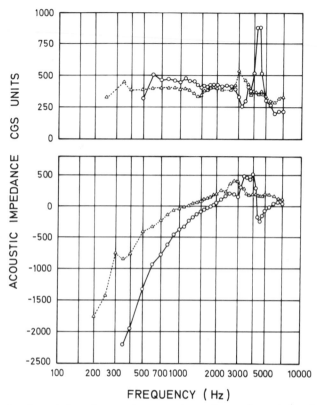

FIGURE 1.27. Acoustic impedance at the tympanic membrane of a cat before (circles and solid lines) and after (triangles and dotted line) the middle ear cavity was opened and the bony septum removed. The upper graph shows the resistive component of the impedance and the lower graph is the reactive component. (From Møller, 1965.)

functions as a stretched membrane, distention will increase in stiffness. The effect is likely to result from the pressure difference between the two sides of the tympanic membrane. The effect of absolute pressure in the middle ear cavity (as long as it is the same on both sides of the tympanic membrane) may be considered negligible since the same constant force is exerted on both the oval and round windows.

Animal experiments have shown that a negative pressure in the middle ear cavity causes a greater decrease in the transmission property of the middle ear than does an equal value of positive pressure (Møller, 1965). Figure 1.28 compares the change in transmission of sound through the

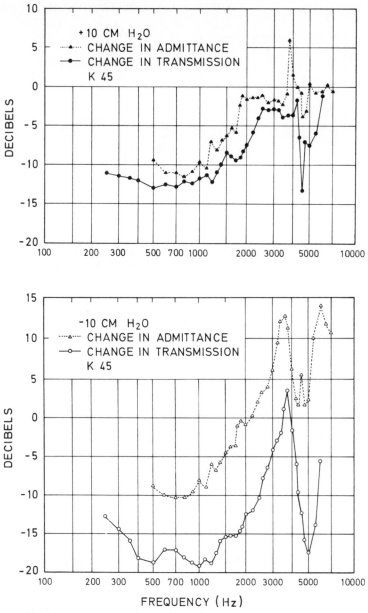

FIGURE 1.28. Change in the acoustic admittance of the tympanic membrane (triangles) and in the transmission (circles) as a result of a + 10 cm H_2O (upper graph) and a − 10 cm H_2O (lower graph) air pressure in the middle ear cavity. Results were obtained from an anesthetized cat. (From Møller, 1965.)

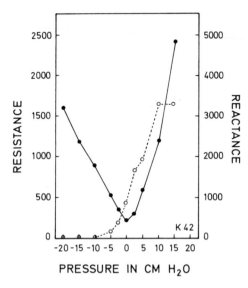

FIGURE 1.29. Acoustic impedance at the tympanic membrane of a cat measured at 1000 Hz as a function of the static air pressure in the middle ear cavity. The two components of the impedance are shown separately. Resistance is indicated by open circles and reactance by filled circles. Results were obtained from an anesthetized cat. (From Møller, 1965.)

middle ear with the change in the ear's acoustic admittance for different frequencies. The upper graph shows the changes for a positive pressure in the middle ear cavities of a cat, and the lower graph gives similar data for a negative pressure. In both cases, the change in admittance and transmission is largest for low frequencies. The change in transmission is larger than the change in acoustic admittance when a negative pressure is present in the middle ear, whereas the transmission and admittance are affected approximately equally by positive pressure. This difference is illustrated in Figure 1.29, where the changes with resistance and reactance are shown separately as a function of the pressure in the middle ear cavity measured in a cat at 1000 Hz. The difference in effect is probably due to a partial functional decoupling of the middle ear from the cochlea for negative pressure. Decoupling probably takes place at the incudostapedial joint. This assumption is based on the finding that a negative pressure in the middle ear results in a marked decrease in the resistive component of the ear's impedance shown in Figure 1.29 (Møller, 1965).

That may explain why there is larger decrease in transmission than in the acoustic admittance for negative pressure, whereas these two measures change almost equally for positive pressure in the middle ear cavity. Data shown in Figure 1.30, obtained in an experiment in a cat

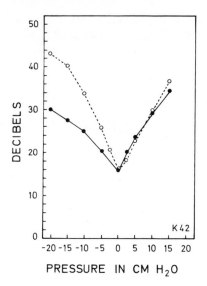

FIGURE 1.30. The change in the acoustic impedance as a result of a change in the air pressure in the middle ear cavity (filled circles) compared with the change in the inverse transmission (open circles). The results are from the same experiment shown in Figure 1.29. The impedance is given in decibels (re 100 cgs units) and the transmission is given in decibels with an arbitrary reference. (From Møller, 1965.)

(Møller, 1965), show both the change in admittance at the tympanic membrane and the change in transmission through the middle ear.

The fact that the ear's acoustic impedance has its lowest value when the air pressure on the two sides of the tympanic membrane has the same value (cf. Figure 1.29) is used in clinical diagnosis of disorders in the middle ear (tympanometry). If there is negative pressure in the middle ear cavity, equal negative pressure in the ear canal restores the acoustic impedance of the ear to its normal low value. Recording of acoustic impedance as a function of the pressure in the ear canal thus provides an easy and noninvasive method of determining if air pressure in the middle ear cavity differs from the ambient pressure and determines the precise value of the pressure.

Acoustic Middle Ear Reflex

ANATOMY

In the middle ear, there are two small muscles called the tensor tympani and the stapedius. The tensor tympani attaches to the manubrium of the malleus, and when it contracts, it draws the tympanic membrane medially. The stapedius muscle is attached to the stapes and draws that bone perpendicularly to its pistonlike movement. The tensor tympani is innervated by the fifth cranial nerve, or N. trigeminus, whereas the stapedius is innervated by the seventh cranial nerve, or N. facialis. The reflex arc is illustrated in Figure 1.31. The pathway of the stapedius reflex in the rabbit is chiefly a chain of four neurons with a small ipsilateral three-neuron link.

The four neurons are (a) primary auditory afferents; (b) neurons of the ventral cochlear nucleus terminating both in the ipsilateral stapedius muscle motoneurons and, presumably, in the neurons of the medial superior olive of both sides; (c) neurons in or near the medial superior olive, terminating in the ipsilateral and contralateral facial motor nucleus; and (d) motoneurons chiefly in the medial part of the facial motor nucleus (Borg, 1973a, b). The tensor tympani is controlled via a pathway involving the medial superior olive or the ventral nucleus of the lateral lemniscus. In addition to these direct connections, there are parallel multisynaptic pathways, including the reticular formation of the brainstem or the extrapyramidal pathway such as the rubrobulbar tract. The details of these pathways are unknown.

PHYSIOLOGY

Contraction of one or both of the two middle ear muscles occurs as an acoustic reflex in response to a loud sound. In man, only the stapedius muscle participates in the acoustic middle ear reflex, whereas both muscles in animals frequently used in auditory research such as cats, rabbits, and guinea pigs can usually be activated by sound. In these animals, the contraction threshold of the tensor tympani is generally higher than that of the stapedius muscle. Contraction occurs in both ears even if the sound reaches only one ear. The contraction in the contralateral ear,

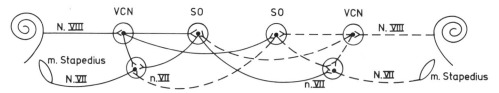

FIGURE 1.31. Reflex arc of the acoustic middle ear reflex: N. VIII, auditory nerve; N. VII, facial nerve; VCN, ventral cochlear nucleus; SO, superior olive; n. VII, facial nerve motoneuron.

however, is weaker, and its threshold is higher than that in the ipsilateral ear (Møller, 1961b). Simultaneous stimulation of both ears produces a stronger response than ipsilateral stimulation does alone (Møller, 1962a).

A convenient measure of the contraction of the middle ear muscles can be made by recording the change in the ear's acoustic impedance. Such measurements may be performed in human subjects as well as in experimental animals. In animals, other methods of measuring acoustic impedance include electromyography (EMG) of the middle ear muscles, and measurement of the change in cochlear microphonic potentials (see p. 89), and directly measuring the mechanical events in the muscles by using suitable mechanical transducers. (For a review of different methods see Møller, 1972, 1974).

Figure 1.32 shows how and intense contraction of the stapedius muscle, elicited by a contralateral pure tone, changes the acoustic impedance of the ear in a human subject with normal ears. The absolute value of the impedance is shown separately (upper two curves) from the phase angle (lower two curves). Impedance increases as a result of middle ear muscle contraction at frequencies below 800 Hz, and it decreases at frequencies above 800 Hz. The reason is that the middle ear is dominated by stiffness below its main resonance frequency (about 800 Hz) and by mass above 800 Hz. Since a contraction of the stapedius muscle mainly increases the stiffness, it shifts the resonance frequency upward. An increase in stiffness above the resonance frequency will result in a decrease in impedance. The change in impedance above 1500 Hz is small. The phase angle shows an increase for frequencies below 1400 Hz.

All of the methods that have been used to record the response of the acoustic middle ear reflex demonstrate that the strength of contraction of the middle ear muscles increases as a function of the sound level eliciting it. Figure 1.33 shows the results of recording the acoustic middle ear reflex in a human subject with normal hearing. The reflex was represented by a change in the ear's acoustic impedance measured at 800 Hz. Impedance change was recorded in both ears simultaneously when either ear was stimulated with sound. The impedance was balanced out before the middle ear muscles were brought to contraction, the response amplitude was normalized to show the same deflection for maximal obtainable change, and the scale shows the relative impedance change as a percent of maximal obtainable change. The response is therefore a result of change in both magnitude and phase of the impedance.

The two left columns in Figure 1.33 show the response when only one ear was stimulated and the right column represents stimulation of both

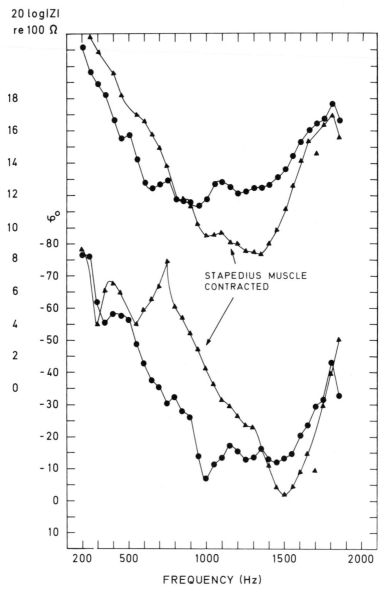

FIGURE 1.32. Change in the acoustic impedance of the ear in a human subject with normal hearing. The upper two curves show the absolute value of the acoustic impedance (in decibels re 100 cgs units) before (circles) and during (triangles) stimulation with a strong tone in the contralateral ear (500 Hz 125 dB SPL). The lower two curves show the phase angle (in degrees) with and without contraction of the middle ear muscles. (From Møller, 1961a.)

42

STIMULATION ON:

LEFT EAR   RIGHT EAR   BOTH EARS

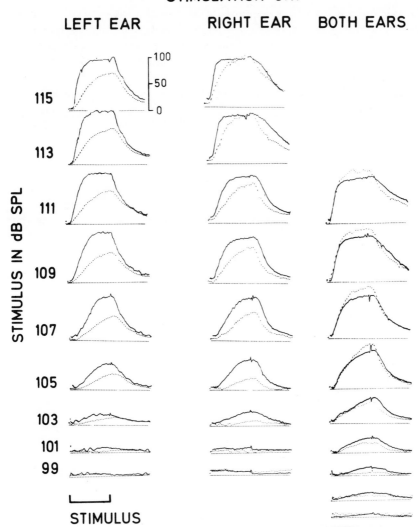

STIMULUS IN dB SPL

115
113
111
109
107
105
103
101
99

STIMULUS

FIGURE 1.33. The change in the acoustic impedance of a human ear as a function of time. The change was recorded simultaneously in both ears in response to stimulation of either left or right ear (monaural) and for stimulation of both ears simultaneously (bilateral). In the two left columns, the solid lines are ipsilateral responses and the dotted lines are the contralateral responses. In the right column (bilateral stimulation) the dotted lines denote the left ear. The stimulus was a 1450 Hz pure tone presented in 500-msec-long bursts. (From Møller, 1962a.)

ears simultaneously. The response increases with increasing sound intensity, and the ipsilateral response is larger than the contralateral response.

Acoustic middle ear reflex, measured as a change in the ear's acoustic
impedance, is displayed quantitatively in Figure 1.34 as a function of
stimulus intensity for ipsilateral and contralateral stimulation. The
relative amplitude of the impedance change (in percent of the maximally
obtainable change) in response to a 500-msec-long tone burst was
measured immediately before the termination of the tone burst. Two
determinations of the reflex amplitude were made at each sound intensity,
one when stimulus intensity was increased from below threshold and one
when it was decreased from maximal response towards threshold values.
Figure 1.34 also shows that the responses near the reflex threshold have a
larger variation than those above threshold (Møller, 1962b). This has also
been shown to be true in animal experiments (Simmons, 1959). Thus, the
actual threshold cannot be determined with certainty, but the intensity

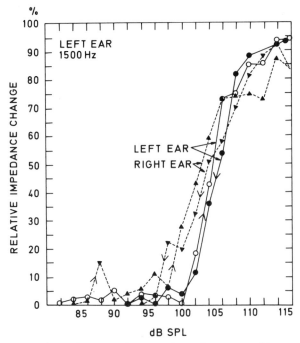

FIGURE 1.34. Impedance change as a function of stimulus intensity. The sound intensity is
first raised from below threshold (in 2-dB steps) until maximal response has been obtained
and then decreased again with the same steps to below threshold. (From Møller, 1962b.)

necessary to obtain a certain small value of impedance change (for example, 10%) is highly reproducible. It is seen from Figure 1.34 that the change in acoustic impedance increases as a function of stimulus intensity for both ipsilateral and contralateral stimulation, to a level where the impedance change reaches a saturation level.

Recorded in that way, response of the acoustic middle ear reflex has a high degree of reproducibility. In fact, it is of the same magnitude as that with which the sound pressure level of the stimulus tone can be determined. Figure 1.35 shows results obtained in the same subject at two different times and at 2 months apart. In this graph, the mean value of the two determinations of the impedance change is shown for intensities in 2-dB steps. Each curve represents the results of one test.

Figure 1.36 shows the relative amplitude of the acoustic impedance

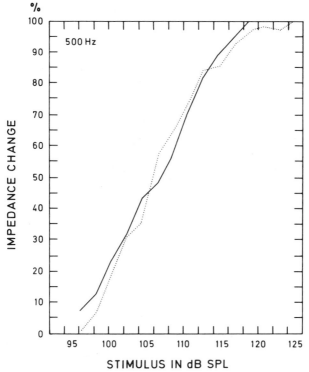

FIGURE 1.35. The response of the acoustic middle ear reflex in a subject with normal hearing measured on two occasions separated by two months. Each data point represents the mean value of two determinations as illustrated in Figure 1.34. (From Møller, 1962b.)

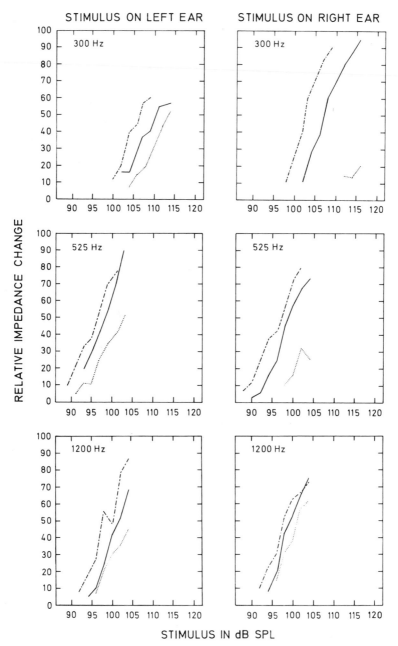

STIMULUS ON LEFT EAR   STIMULUS ON RIGHT EAR

300 Hz

300 Hz

525 Hz

525 Hz

1200 Hz

1200 Hz

RELATIVE IMPEDANCE CHANGE

STIMULUS IN dB SPL

—— IPSILATERAL EAR
·········· CONTRALATERAL EAR
BOTH EARS STIMULATED
–·–·EQUAL INTENSITY

46

change in a typical, normal-hearing subject for ipsilateral, contralateral, and bilateral stimulation. Results from these different frequencies are shown. The measurements were performed as illustrated in Figure 1.34. Impedance change was determined at stimulus levels with 2-dB intervals in between and the mean value of two determinations at each intensity level are shown. Curves showing the response to ipsilateral and bilateral stimulation are nearly parallel, indicating that the reflex is about 3 dB more sensitive in response to bilateral stimulation than it is to ipsilateral stimulation. To compensate for the small differences in the sensitivity of the two ears, bilateral stimulation was arranged so that the sound level in the two ears was matched to yield approximately the same ipsilateral response when presented alone.

When stimuli of the same sound intensity are used, the response to contralateral stimulation is smaller than that to ipsilateral stimulation. The maximally obtainable impedance change is often smaller for contralateral stimulation than it is for ipsilateral stimulation. The difference in sensitivity to contralateral-ipsilateral stimulation varies from person to person from 2 to 14dB (Møller, 1962a). The difference in sensitivity between ipsilateral and contralateral stimulation is different for the different frequencies of the tone stimuli used to elicit the reflex response.

It is worth noting that the natural stimulation of the middle ear reflex is bilateral. It is therefore unfortunate that most clinical tests of the acoustic middle ear reflex are performed using contralateral stimulation.

Sensitivity of the Reflex to Pure Tones

In man the contralateral "threshold" of the reflex in response to pure tone frequencies between 200 and 4000 Hz (Figure 1.37), is nearly parallel to the threshold of hearing, the difference being about 80 dB (Møller, 1962b). Results in that graph were obtained using tone bursts of 500-msec

---

FIGURE 1.36. Stimulus response curves for the acoustic middle ear reflex in a typical human subject with normal hearing. The curves show the change in the ear's acoustic impedance as a function of the intensity of the sound stimulation. The left-hand graphs represent stimulation of the left ear and the right-hand graphs show responses to stimulation of the right ear. The individual graphs represent different stimulation frequencies. Solid lines indicate the impedance recorded in the ear from which the reflex is elicited; dotted lines show the contralateral response; dashed lines show impedance change in response to bilateral stimulation. The duration of the stimuli in all cases was 500 msec. The impedance change is given at 2-dB intervals, and each data point represents the mean value of two determinations, one when the intensity was increased and one when it was decreased, as illustrated in Figure 1.34. The sound level given is the sound pressure level near tympanic membrane measured with a probe microphone during the actual experiment. (From Møller, 1962a.)

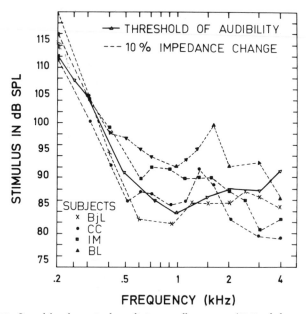

FIGURE 1.37. Sound level required to elicit a small response (10% of the maximal obtainable response) of the middle ear reflex through contralateral stimulation with pure tones. The individual curves are from individual subjects. They were young people with normal hearing. For comparison, the threshold of audibility curve is depicted elevated 80 dB. (From Møller, 1962b.)

duration as stimuli. The sound level (in decibels, SPL) necessary to elicit 10% of the maximum impedance change is given as "threshold." As just mentioned, this threshold value is highly reproducible, whereas absolute threshold values show large variations on different occasions in the same subject. In Figure 1.37, there is great individual variation in this 10% "threshold," but the curves showing relative threshold of the reflex are generally parallel to the curve showing threshold of the reflex of audibility elevated about 80 dB (80-dB hearing level). It is interesting to note that the acoustic middle ear reflex threshold does not follow the equal-loudness curve at 80 dB, but rather it follows the threshold curve displaced 80 dB (Møller, 1962b).

Sensitivity to Different Types of Stimuli

In general, the sensitivity of the reflex decreases as the duration of the stimulus sound is decreased below about 200 msec. For longer sounds, the sensitivity of the reflex is independent of the duration. (Djupesland &

Zwislocki, 1971). Stimulus response curves obtained using tone bursts with durations shorter than 200 msec have a different shape than those obtained with durations longer than 200 msec. The maximal, or saturation, value of the reflex response is lower for sounds with a short duration (Møller, 1962a). Figure 1.38 illustrates the stimulus response relationship for 500-msec stimulation and 25-msec stimulation. The slope of the stimulus responses curves for the short sound is much smaller than it is for long-duration sounds. Also, the difference in the shapes of the stimulus response curves for these two types of sound is more pronounced for the response to contralateral stimulation than to ipsilateral and bilateral stimulation. This is important to note, since most of the work concerning the function of the middle ear reflex is based on contralateral recording of the reflex response. There is, however, a large individual variation.

Results shown previously have been based upon pure tone stimuli. Such sounds are different from normal sounds that usually have energy distributed over a large frequency range. The acoustic middle ear reflex response to such natural sounds is different in many ways, although the experimental data are limited. As an example, Figure 1.39 illustrates how the sensitivity for the middle ear reflex differs between pure tones and bandpass-filtered noise. The difference in sound level required to obtain

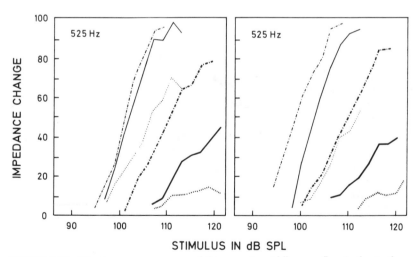

FIGURE 1.38. Stimulus response curves of the acoustic middle ear reflex similar to those seen in Figure 1.36 but for two different durations of the stimulus tone: 500 msec, thin lines; and 25 msec, heavy lines. Bilateral stimulation: dashes and dots; ipsilateral stimulation: solid lines; contralateral stimulation: dots. The results are from a subject with normal hearing. (From Møller, 1962a.)

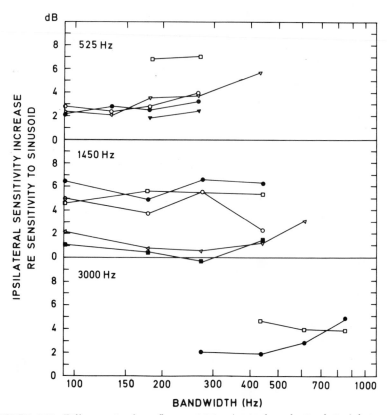

FIGURE 1.39. Difference in the reflex sensitivity (to ipsilateral stimulation) between sinusoidal stimulation and bandpass-filtered noise. The frequency of the tones is the same as the center frequency of the noise. The difference in sensitivity shown is the difference in sound level required to evoke a response that is 10% of the maximally obtainable response. These values are shown as a function of bandwidth of the noise for three different center frequencies. The different curves refer to different subjects, all of whom had normal hearing. (From Møller, 1962a.)

an ipsilateral response of 10% of the maximal obtainable impedance change is plotted as a function of the bandwidth of the noise for these different center frequencies (Møller, 1962a). Sensitivity to noise is generally higher than to a pure tone of the same physical intensity, and the sensitivity increases with increased bandwidth. A more systematic study of the realtions between reflex threshold and noise bandwidth was presented by Djupesland and Zwislocki (1973) and Popelka, Margolis, and Wiley (1976) using the contralateral reflex. They found that using a concept similar to that of the critical band used to describe masking, they could describe the average change in sensitivity of the reflex as a function of bandwidth. The value of the critical band obtained for the acoustic middle ear reflex was, however, much larger than the psychoacoustic critical band value. Studies by Djupesland and Zwislocki (1973) and Popelka *et al.* (1976) on responses to complex sound were performed using contralateral stimulation. The sensitivity of the reflex to complex sounds presented bilaterally cannot necessarily be predicted on the basis of results using contralateral stimulation. That is supported by the finding that the difference between the response to bilateral, ipsilateral, and particularly to contralateral stimulation, depends on the repetition frequencies when clicks are used to elicit the reflex response. This is illustrated in Figure 1.40, which shows stimulus response curves similar to those in Figure 1.36 for bandpass-filtered clicks with different repetition rates (Møller, 1962a).

Thus, the sensitivity of the middle ear reflex in response to bilateral stimuli and to broadband sounds is considerably higher than it is in the common test situation using contralateral stimulation with pure tones. If the average difference between bilateral and contralateral stimulation is assumed to be about 10 dB and the difference in senstivity between pure tones and natural sounds is taken to be 4 dB, then the *natural* threshold of the acoustic middle ear reflex is not 80–85 dB above hearing threshold but rather 65–70 dB, which makes it likely that this reflex is activated in many normal listening situations.

Temporal Characteristics of the Reflex

Muscle reflex of the middle ear is rather slow. Its latency in man decreases with increasing stimulus intensity from above 150 msec near threshold to about 35 msec near maximal contraction (Møller, 1958). The build-up of the contraction in response to a tone burst is slower at low-stimulus intensities, where it may take more than 500 msec to reach a steady-state value. At high intensities it takes about 150 msec to reach a steady-state value.

The temporal pattern of the reflex response in a human subject is shown in detail in Figure 1.41, where the ipsilateral response to tone bursts of 500 and 25 msec are shown for 525-Hz tones and 1450-Hz tones. Response to the two frequencies (525 and 1450 Hz) differs when the tones are presented in 500-msec bursts, but the responses to 25-msec tone bursts are similar. This is because the sound terminates before the reflex response

STIMULATION WITH 525Hz SINE-WAVE BURST

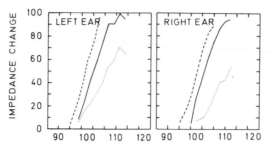

STIMULATION WITH BANDPASS-FILTERED PULSES F₀-525Hz B-100Hz

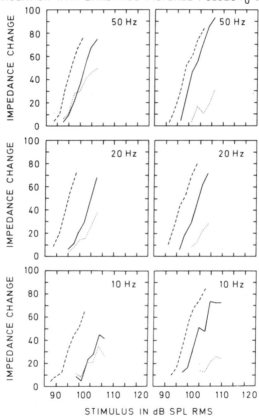

STIMULUS IN dB SPL RMS

FIGURE 1.40. Stimulus response curves similar to those seen in Figure 1.36 comparing the response to pure tones and bandpass-filtered repetitive clicks of different repetition rates. The responses to bilateral, ipsilateral, and contralateral stimulation are shown using the same symbols as in Figure 1.36. All the sounds were presented in 500-msec-long bursts. (From Møller, 1962a.)

52

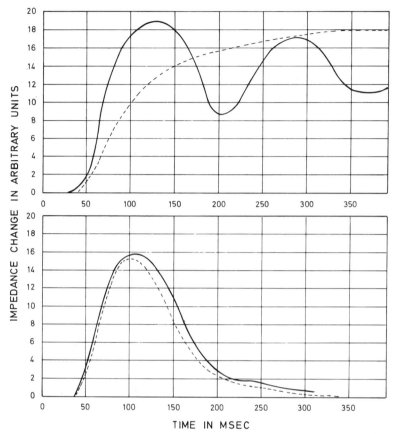

FIGURE 1.41. Ipsilateral reflex response (shown as impedance change as a function of time) to a 500-msec tone burst (upper graph) and a 25-msec tone burst (lower graph). The time scale is referred to the beginning of the tone burst. Solid lines are the response to a 525-Hz tone and dashed lines are the responses to a 1450-Hz tone. The rise times of the tone bursts were about 5 msec, and the response time of the impedance measuring device was about 10 msec. (From Møller, 1962a.)

(muscle contraction) can alter the transmission through the middle ear. The damped oscillation shown in response to the 525-Hz, 500-msec tone burst is a result of the regulation of sound transmission through the middle ear by contraction of the stapedius muscle. Response to a tone burst at the low frequency of 525 Hz, at which the reflex exerts a great influence on sound transmission through the middle ear, is compared to the response to a tone at 1450 Hz where this regulatory influence is negligible.

Further analysis indicates that the onset response of the middle ear reflex is approximately twice as fast as the offset response to a tone burst of 500 msec or longer (Møller, 1962a). Whereas the latency buildup and decay times are sufficient to describe the dynamic properties regarding tone bursts of near-abrupt onset and offset, these factors fail to describe the reflex response to natural sound, the intensity of which varies in an irregular manner. The nonlinear nature of the middle ear reflex prevents a general description of its dynamic properties, but for small changes in stimulus intensity, these are known to vary as a function of stimulus intensity (Borg, 1971). Figure 1.42 shows the response of the middle ear reflex to a stepwise increase in stimulus intensity. The different curves represent different levels of constant stimulation from which excitation is increased.

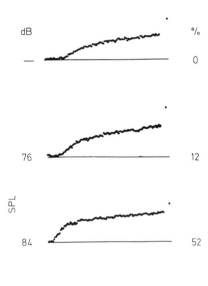
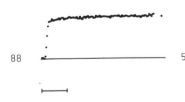

FIGURE 1.42. Response of the acoustic middle ear reflex to ipsilateral stimulation when the reflex is constantly elicited by contralateral stimulation at various levels (indicated by legend numbers). The results were obtained from an unrestrained rabbit using 2000-Hz pure tone stimulation. (From Borg, 1971.)

Because of the slow buildup of the contraction of the middle ear muscles, the reflex does not effect rapid changes in sound intensity but exerts a slow regulatory influence on the transmission of sound to the inner ear. Rapid changes in intensity are thereby transmitted with little or no attenuation by the reflex. The result is that the reflex *suppresses* the transmission of slow changes in sound intensity to the cochlea, thus *compressing* the amplitude range for steady sounds without reducing that for rapid changes in sound level (see Chapter 4 on adaptation).

The effect of the middle ear muscle (stapedius) contraction on sound transmission through the middle ear is different for different frequencies, as will be later described in detail (p. 58). Since the reflex is a feedback system, that fact is also manifest in the temporal response of the reflex. Reflex response to a low-frequency sound thus has a faster onset than that to a high-frequency sound, and has a damped oscillation of its onset because the contraction of the stapedius reflex has a greater influence on middle ear transmission at low frequencies than it does at high frequencies. Reflex thus regulates sound transmission to the cochlea to a greater extent for frequencies below the principal resonance frequency of the middle ear than it does for frequencies above the resonance frequency of the middle ear (about 1000 Hz). Above approximately 1000 Hz, the attenuating effect of a contraction of the stapedius muscle becomes small, and consequently above that frequency, the acoustic middle ear reflex does not function as an effective regulatory mechanism. That can be seen from the time course of the response in Figure 1.43, which shows the ipsilateral response to tones of different frequencies. Above 1000 Hz there is no oscillation seen in the response, indicating that the regulative power of the reflex is low above that frequency (Møller, 1962a).

Effect of Drugs

Drugs, such as barbiturates and ethanol, decrease the sensitivity of the middle ear reflex (Borg & Møller, 1967; Figure 1.44). Figure 1.44 shows that response to the same sound intensity is smaller after ethanol is ingested. The time course of the reflex response also changes as a result of ethanol ingestion. Ipsilateral and contralateral responses are affected almost to the same degree in the same person (Borg & Møller, 1967), but there is great variation among individuals. Figure 1.45 shows the difference in sound intensity necessary to evoke a response that is 10% of the maximal obtainable response. The mean value of the elevation in threshold is shown in Figure 1.46 for the ipsilateral and contralateral

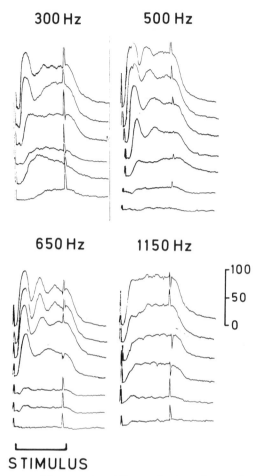

FIGURE 1.43. Response of the acoustic middle ear reflex in a subject with normal hearing to tones of four different frequencies. The tones were presented in bursts of 500-msec duration. The scale to the right indicates the impedance change (ipsilateral) as a percentage of maximal obtainable change. (From Møller, 1962a.)

responses to 500- and 1450-Hz tones as a function of blood alcohol level (The sound intensity necessary to evoke a 10% impedance change is discussed on p. 45). An alcohol level of .05% in the blood increases the threshold by 2 dB on average, with a great individual spread. An alcohol level of .1% in the blood increases the threshold by 5 dB, and an alcohol level of .14% increases the threshold by 13 dB at 500 Hz. Barbiturates also

BEFORE   AFTER
ETHANOL   ETHANOL

A.T. 13.11.65

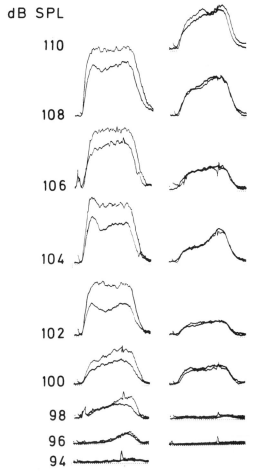

dB SPL

110

108

106

104

102

100

98

96

94

FIGURE 1.44. Effect of ethanol on the response of the acoustic middle ear reflex in a human subject with normal hearing and normal middle ear reflex. The reflex response to 500-msec tone bursts is shown for the ipsilateral (solid line) and contralateral ear (broken line). The stimulus intensity is given by legend numbers. The right-hand recordings were made 30 minutes after ingestion of 105 ml alcohol and the blood alcohol level was .12% at the time of recording. (From Borg & Møller, 1967.)

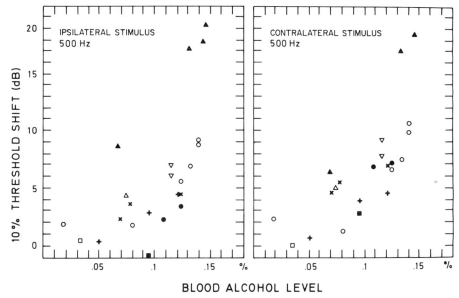

FIGURE 1.45. Effect of ethanol on the sensitivity of the acoustic middle ear reflex. The difference in sound level necessary to elicit a response that is 10% of the maximally obtainable response is shown as a function of blood alcohol level. The different symbols show individual results. The results are based on studies performed in nine experimental subjects. (From Borg & Møller, 1967.)

have a depressive effect on the acoustic middle ear reflex in man. The effect is greater for the contralateral ear than for the ipsilateral. The slope of the stimulus response curve decreases, more so for the contralateral than for the ipsilateral ear, thus indicating a larger effect above threshold (Borg & Møller, 1967). These results were supported by experiments in rabbits that showed similar effects of barbiturates (Borg & Møller, 1975).

Effect of Middle Ear Muscle Contraction on Middle Ear Transmission

Contraction of the middle ear muscles results in reduced transmission of sound through the middle ear. Consequently, the middle ear reflex functions as a control system and tends to keep the input at the cochlea independent of variations in the sound level. In man, a 1-dB increase in sound pressure at the tympanic membrane, as long as the frequency was below the middle ear's natural frequency (Borg, 1968), has been shown to give rise only to a .7-dB increase in the sound energy reaching the cochlea.

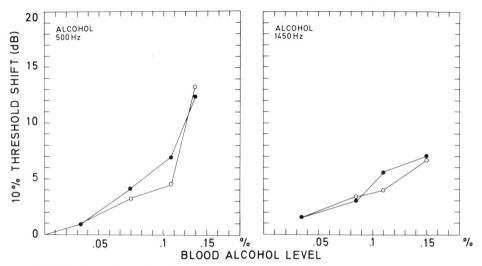

FIGURE 1.46. Mean values of the increase in stimulus intensity necessary to obtain an un-changed reflex response (of 10% of the maximally obtainable response) as a function of blood alcohol concentration. Open circles represent the uncrossed reflex and filled circles the crossed reflex. The results are based on 22 tests in nine subjects. (From Borg & Møller, 1967.)

At higher frequencies, the effect of low contractile force of the stapedius muscle is minimal, but near maximal contraction transmission undergoes a substantial reduction, even at frequencies above the main resonance of the middle ear.

Reduction in transmission brought about by contraction of the stapedius muscle at low frequencies is likely to result from increased stiffness of the middle ear. At high frequencies, attenuation is probably caused by a change in the vibratory mode of the stapes. It is also assumed that the contraction of the stapedius muscle changes the vibration of the footplate of the stapes from a pistonlike motion to a rolling motion (Møller, 1961a). The latter motion is less efficient in setting the cochlear fluid into motion; therefore, the transmission of sound is reduced through the middle ear.

It should be noted that the reflex response elicited by a high-frequency component of a complex sound may affect the transmission of low-frequency components of that sound more strongly than it will the high-frequency component that elicited the response. This phenomenon is independent of the intensity of the low-frequency sounds, which may, in fact, be below the level where they would elicit a reflex response.

The effect of contraction of the stapedius muscles has been measured in animal experiments, and Figure 1.47 shows the change in transmission caused by contraction of the stapedius muscle in the cat. The largest change occurs at low frequencies, and above 1500 Hz there is even a small increase in transmission. Also shown is the change in acoustic admittance at the tympanic membrane. As discussed earlier, the change in admittance and transmission would be equal if the effect of the shunting elements in the analog circuit of the middle ear was negligible. As shown in Figure 1.47, there is a certain difference between the two curves, which indicates that the influences of shunting elements are not large but not negligible.

Difference in Effect of the Two Muscles

Experiments on cats and rabbits have shown, as expected, that the tensor tympani muscle draws the tympanic membrane inward. This has been shown by stimulating the tensor tympani muscle electrically in animal experiments. Results of similar experiments on the stapedius muscle reveal

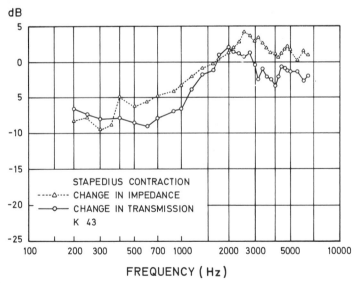

FIGURE 1.47. Change in transmission (solid line) compared with the inverse change in the acoustic impedance (change in acoustic admittance) at the tympanic membrane during contraction of the stapedius muscle. Results were obtained from a cat. (From Møller, 1965.)

that when contracting it produces very little or no movement of the tympanic membrane (Møller, 1964b). Figure 1.48 shows results of such experiments. Displacement of the tympanic membrane (upper row) is shown together with the change in acoustic impedance and in the cochlear microphonic potential (lower rows of curves). The left most curves represent contraction of the tensor tympani; the middle, the stapedius muscle; and the rightmost curves represent the combined effect of the two muscles. The two muscles have nearly the same effect on acoustic impedance and cochlear microphonics (CM), the latter representing the change in transmission through the middle ear. The stapedius muscle is faster than the tensor tympani. When both muscles are brought to contraction together, they affect the impedance and the CM to a higher degree than when each of the two muscles is brought to contraction separately. The two muscles act differently in regard to movement of the tympanic membrane. The tensor tympani moves the tympanic membrane inward when contracting (upward deflection in Figure 1.48), whereas the

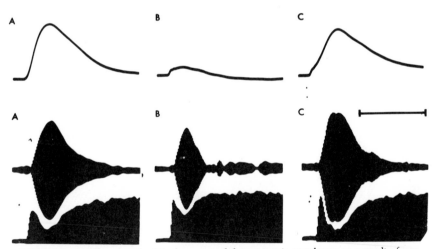

FIGURE 1.48. *Upper graphs:* Displacement of the tympanic membrane as a result of contraction of the tensor tympani muscle (A), stapedius muscle (B), and both together (C). Results are from an experiment in an anesthetized cat and the muscle contractions were brought about by a single electrical shock applied to the muscle. The displacement of the tympanic membrane was measured by recording the change in air pressure in the sealed ear canal. *Lower graph:* Changes in acoustic impedance (upper row of recordings) and change in the cochlear microphonic potential when the middle ear muscles were brought to contraction in the same way as in the upper recordings. The change in the acoustic impedance was measured at a frequency of 800 Hz. Horizontal calibrations: 100 msec; maximal pressure: approximately 0.2 mm H_2O. (From Møller, 1964b.)

stapedius muscle does not move the tympanic membrane practically at all. When brought to contraction simultaneously, the two muscles produce less movement of the tympanic membrane than the tensor tympani muscle does alone. The stapedius muscle thus impedes the tensor tympani in moving the tympanic membrane. These results were obtained from experiments performed on cats, and similar results were obtained from the rabbit.

Figure 1.49 shows the displacement of the tympanic membrane and the change in acoustic impedance when the two muscles were brought to contraction by electrical stimulation. The graphs B and D represent different times between eliciting the contraction in the two muscles. In B, contrac-

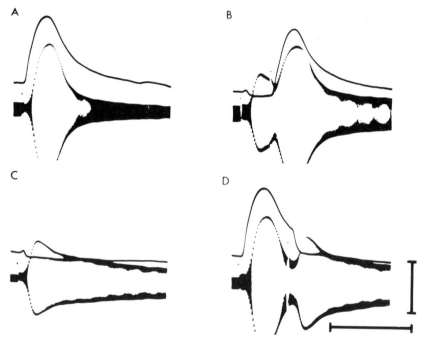

FIGURE 1.49. Displacement of the tympanic membrane (solid line) and change in acoustic impedance measured in an experiment similar to that shown in Figure 1.48. In (A) only the tensor tympani contraction was elicited, and in (C) is shown the response of the stapedius muscle alone. The muscles were brought to contraction by electrical shocks. In (B) the shock applied to the stapedius muscle was presented a short interval before the one applied to the tensor tympani. The response shown in (D) was obtained when the shock to the tensor tympani was applied prior to that to the stapedius muscle. Horizontal calibration: 100 msec; vertical calibration: 0.2 mm H_2O.

tion of the stapedius muscle was elicited before the tensor tympani was stimulated. In D, a contraction of the stapedius muscle was elicited after the tensor tympani was stimulated. Regarding acoustic impedance and sound transmission through the middle ear, the two muscles exert fairly equal effects, and their combined effect is somewhat greater than that produced by either of the two alone (Møller, 1964b).

In conclusion, the two muscles act synergistically with regard to changing the acoustic properties of the middle ear (impedance and transmission), whereas the stapedius works antagonistically to the tensor tympani in displacing the tympanic membrane.

Effect of Middle Ear Reflex on Hearing

The fact that, at least at low and moderate contraction levels, the middle ear reflex influences low frequencies more than the middle and high frequencies, may be of importance in reducing the masking effect of low-frequency sounds on high-frequency sounds. The function of the middle ear reflex in protecting the ear against overstimulation has been studied in patients suffering from acute Bell's palsy with total unilateral paralysis of the stapedius muscles (Zakrisson, 1975). In such patients, there is a higher tempory threshold shift in the affected ear after exposure to low-frequency noise than in an ear with a normal stapedius reflex. Such patients also exhibit decreased discrimination of speech in the affected ear when the sound level of speech is raised above about 90 dB, whereas the intelligibility of similar speech is unimpaired in the normal ear (Borg & Zakrisson, 1973).

Figure 1.50 shows speech intelligibility as a function of sound intensity for two persons—one with normal middle ear reflexes and one without one middle ear reflex due to stapedius paralysis. Speech intelligibility for the person without the stapedius reflex falls rapdily above 90 dB whereas it remains almost unchanged for the person with a normal stapedius reflex.

It has been shown that in humans, middle ear muscles are activated by one's own speech to a higher degree than would be expected from the level of sound that actually reaches the ear, and that stapedius muscle activity could occasionally be detected before speech actually commenced (Borg & Zakrisson, 1975). Results were obtained by recording the electrical activity in the stapedius muscle in humans undergoing ear surgery. An example of such a recording is shown in Figure 1.51. This graph shows recordings of both the electrical activity in the muscle (EMG) and the

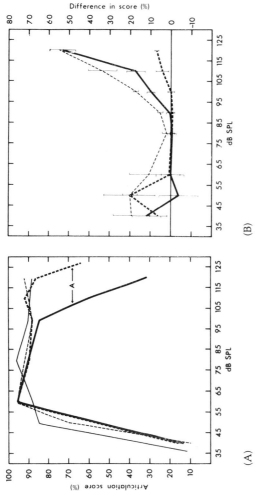

(A)

(B)

FIGURE 1.50. Effect on speech discrimination of stapedius muscle paralysis (during Bell's Palsy). (A) Articulation scores on nonsense monosyllables (in percentage) shown as a function of the maximal speech level (decibels re 20μPa) in a subject during stapedius muscle paralysis (heavy continuous line) and after complete recovery of stapedius muscle function (thin continuous line). The articulation score in the unaffected ear is shown by a heavy broken line. (B) Average difference in articulation scores as a function of sound pressure level of nonsense monosyllables. Heavy continuous line shows the scores obtained in the unaffected ear minus the affected ear (seven subjects). Thin line: Scores in unaffected ears after recovery from paralysis minus scores during paralysis (six subjects). Vertical bars show standard errors of the mean. (From Borg & Zakrisson, 1973.)

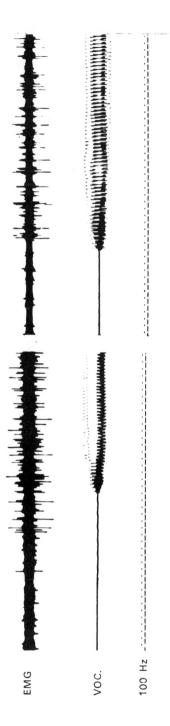

EMG

VOC.

100 Hz

97 dB SPL

94 dB SPL

FIGURE 1.51. Electrical activity (EMG) of the stapedius muscle during vocalization in a human subject (upper trace) shown together with a recording of the sound near the person's mouth (middle trace). The muscle response was recorded during a middle ear operation and the person was asked to vocalize an [a:]. The sound was recorded by a microphone placed 20 cm in front of the mouth. The lower trace shows a 100 Hz calibration signal. (From Borg & Zakrisson, 1975.)

sound recorded by a microphone placed near the mouth of the person. The muscle action potentials are clearly seen to appear prior to the vocalization. This action may be of importance in preserving hearing capacity immediately after speech or as an antimasking mechanism during one's own speech, preventing the low-frequency component of one's speech from masking other high-frequency sounds. In the flying bat, a similar response pattern was seen in connection with the emission of echolocation sounds (Henson, 1965).

The Inner Ear

ANATOMY

Figure 1.52 shows a schematic drawing of the vestibular apparatus and the cochlea of the human inner ear. Only the cochlea will be considered here. The cochlea is snail-like in shape. In man, it has 2¾ turns and its length when unrolled, is about 3.5 cm. The entire cochlea is filled with fluid. It is divided into three longitudinal canals: the scala vestibuli, the scala tympani, and, in the middle, the scala media (Figure 1.53). The scala media is separated from the scala vestibuli by Reissner's membrane and

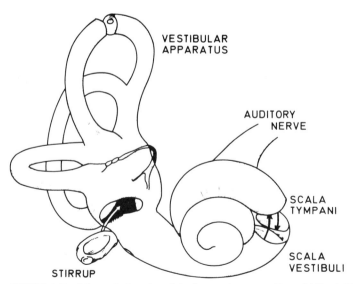

FIGURE 1.52. Schematic drawing of the human inner ear. (From Melloni, 1957.)

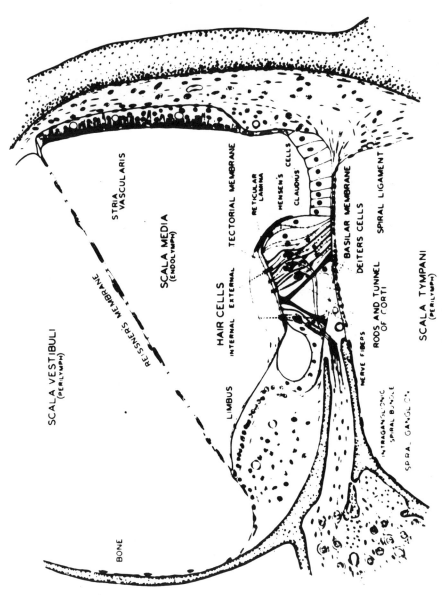

SCALA VESTIBULI
(PERILYMPH)

REISSNER'S MEMBRANE

STRIA VASCULARIS

SCALA MEDIA
(ENDOLYMPH)

TECTORIAL MEMBRANE

RETICULAR LAMINA

HENSEN'S CELLS

CLAUDIUS CELLS

BASILAR MEMBRANE

DEITERS CELLS

SPIRAL LIGAMENT

HAIR CELLS

INTERNAL EXTERNAL

RODS AND TUNNEL OF CORTI

SCALA TYMPANI
(PERILYMPH)

LIMBUS

NERVE FIBERS

INTRAGANGLIONIC SPIRAL BUNDLE

SPIRAL GANGLION

BONE

FIGURE 1.53. Cross-section through one turn of the cochlea. The picture shows the second turn of a guinea pig's cochlea. (From Davis et al., 1953.)

67

from the scala tympani by the basilar membrane. The organ of Corti, which embodies sensory cells (hair cells), lies along the basilar membrane. These sensory cells convert vibrations of the basilar membrane into nerve impulses within individual fibers of the auditory part of the eighth cranial nerve, thus transmitting information about the sound to the brain.

The stapes footplate is located in the oval window, an opening in the bony shelf of the cochlea into the scala vestibuli. The round window, a similar opening covered by a membrane, exists in the scala tympani. When the stapes vibrates, the fluid in the cochlea is set into motion. *In* and *out* movements of the stapes consequently cause the round window to bulge *out* and *in*. If the round window were absent, pressure applied to the stapes would not produce any movement of the fluid due to the incompressiblity of the cochlear fluid and the cochlear walls. The motion of the cochlear fluid, in turn, causes a wave motion along the basilar membrane. That wave motion is closely related to the separation of spectral components of complex sound that occurs in the ear. The acoustic properties of the basilar membrane will be later described in detail. There is a small aperture in the basilar membrane near the apex, the helicotrema, which keeps the pressure within the scala tympani and the scala vestibuli equal.

THE INNER EAR FLUIDS

The scale vestibuli and scala tympani both contain perilymph, a fluid similar in ionic composition to extracellular fluid and, therefore, rich in sodium (Na) and poor in potassium (K). The ionic composition of the fluid of the scala media (endolymph) is radically different because it is similar to that fluid usually found *within* the cells (intracellular fluid; see, e.g., Bosher, 1979). The functional importance of the composition of these fluids is not fully understood, but a change in composition disturbs the normal function of the sensory hair cells.

ANATOMY OF HAIR CELLS

The sensory cells that transform the motion of the basilar membrane into nerve impulses of the individual auditory nerve fibers are the hair cells along the basilar membrane in the organ of Corti. In the mammalian ear, hair cells are organized into one row of inner hair cells and three to five rows of outer hair cells. Figure 1.54 is a scanning electronmicrograph showing the organization of hair cells in the basilar membrane of a

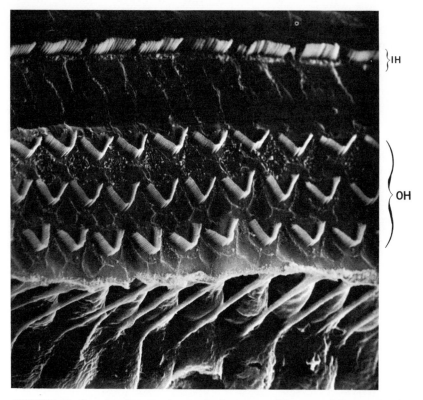

FIGURE 1.54. Scanning electron micrograph of a segment of the organ of Corti in a monkey with the tectorial membrane removed to show the organization of the hair cells. One row of inner hair cells (IH) is shown at the top of the picture and three rows of outer hair cells (OH) are shown in the middle with their typical W-shaped formation of hairs. (Courtesy of Hans Engström, Uppsala.)

monkey. The W-shaped organization of the hairs of the outer hair cells is clearly seen.

Hair cells have evolved from receptor cells of the lateral line organ of fish. The next step in phylogenetic evolution was the vestibular system, where hair cells differentiated into two types. Further changes occurred through the evolution from amphibians to reptiles and birds. The different species of reptiles in particular show a variety of different organizations of hair cells, and it has been said that reptiles constitute "nature's experimental workshop regarding hearing." All hair cells found in vertebrate animals that evolved prior to mammals had both stereocilia and one kinocilium. The hair cells are different in the mammalian cochlea because the kinocilium does not exist in adults. In the fetal stage even the mammalian hair cells have one kinocilium in addition to the stereocilia, thus similar to hair cells of earlier animals (birds, reptiles, and fishes). After birth the kinocilia of cochlear hair cells either disappear or remain as reduced vestigial appendages (Engström & Engström, 1978). (In the vestibular system the kinocilium persists into adult life.)

Stereocilia can be regarded as exvaginations of the cell membrane. They are thus covered by the same cell membrane as the entire hair cell. Within the stereocilia, there are filaments that have been shown to consist of actin, the contractile protein of muscle (Flock & Cheng, 1977; Flock, 1980). Stiffness of the stereocilia may, therefore, be controlled by chemical or electrical influences. Studies have shown that stereocilia are equipped with an electrical surface charge that keeps them separated and also adds to the stiffness of the hairs (Flock et al., 1977). It is usually assumed that it is the mechanical movement or bending of the hairs (stereocilia) that causes excitation of the cell and eventually gives rise to the change in the impulse activity of the nerve fibers that connect to the cells and form the auditory nerve.

Many scientists have studied the linkage of hairs with the tectorial membrane. The tectorial membrane covers hair cells in the manner of a roof hinged on only one side. Hairs of the outer hair cells are embedded in the tectorial membrane, but those of the inner hair cells are probably not (Kimura, 1965; Lim, 1971, 1972). It is generally assumed that the connection between the hairs and the tectorial membrane is of great importance for the transfer of the motion of the basilar membrane into a mechanical deformation of hair cells (bending of the hairs). It has thus been assumed that the tectorial membrane plays an important role in the series of events that occurs in the cochlea and finally results in elicitation of nerve impulses in the fibers of the auditory nerve. Hairs that are not attached to the tectorial membrane should be displaced by movement of the fluid (endolymph) around hairs which then hypothetically constitute the efficient mechanical stimulus of the hair cells (Billone, 1972).

Excitatory synaptic junctions connect hair cells to the afferent (ascending) nerve fibers and to the auditory part of the eighth cranial nerve, whereas inhibitory synaptic junctions connect the efferent (descending) fibers of Rasmussen's bundle (the olivocochlear bundle) to hair cells. Figures 1.55 and 1.56 are schematic illustrations of inner and outer hair cells with their synaptic connections. The inner and outer hair cells differ

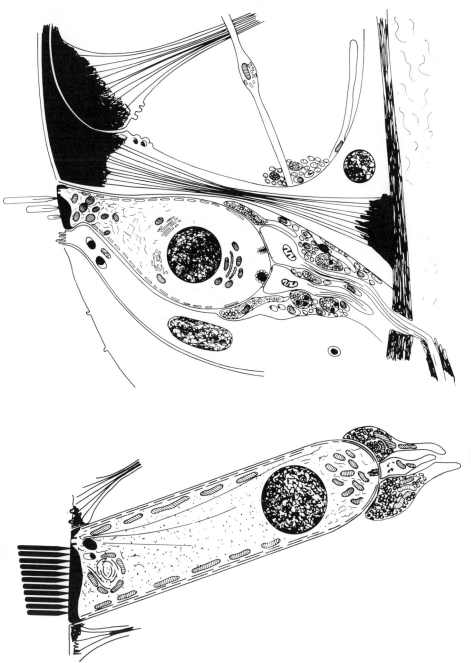

FIGURE 1.55. Schematic picture of outer (left) and inner (right) hair cells showing their synaptic connections. The figure is based on electron microscopic work on hair cell morphology. (Courtesy of Hans Engström, Uppsala.)

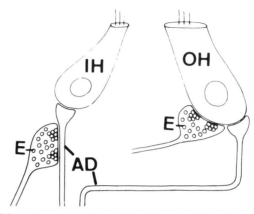

FIGURE 1.56. Schematic representation of the afferent and the efferent synaptic connections to inner (IH) and outer (OH) hair cells in the cat. Efferent endings are indicated by E and
afferent dendrites by A. At the outer hair cells (OH), synaptic contacts are almost exclusively with the sensory cells, whereas at the inner hair cells (IH) the efferent synaptic contact
is with the afferent dendrites. (From Spoendlin, 1970.)

in several ways. Inner hair cells are larger than the outer, and their nuclei
are relatively small, whereas the nucleus of the outer hair cells almost fills
the diameter of the cell (see Figure 1.55). Although inner hair cells are
tightly surrounded by supporting cells, only the lower part of outer hair
cells is fixed by supporting cells. The outer hair cells can then be regarded
as being almost freely moveable along approximately their entire length.
The functional role of the organization into inner and outer hair cells and
of the various differences between them is not known.

Approximately 95% of the afferent fibers of the auditory nerve innervate inner hair cells, whereas the remaining 5% of the afferent fibers terminate on the outer hair cells (Spoendlin, 1970). Innervation of the hair
cells is illustrated schematically in Figures 1.57A and B. Thus, many afferent nerve fibers connect to synapses on inner hair cells. It is different
on the outer hair cells. Here, both efferent and afferent fibers make synaptic connection with the cell, and one afferent nerve fiber connects to many
hair cells. There are approximately 50,000 afferent nerve fibers in the
auditory nerve of a cat. Each of the inner hair cells is then innervated by as
many as 20 afferent fibers, whereas one afferent fiber innervating outer
hair cells takes a long spiral course (about .6–.7 mm) before sending its
terminals over a distance of approximately 200μm to the outer hair cells.
Each fiber sends collaterals to about 10 outer hair cells over that distance

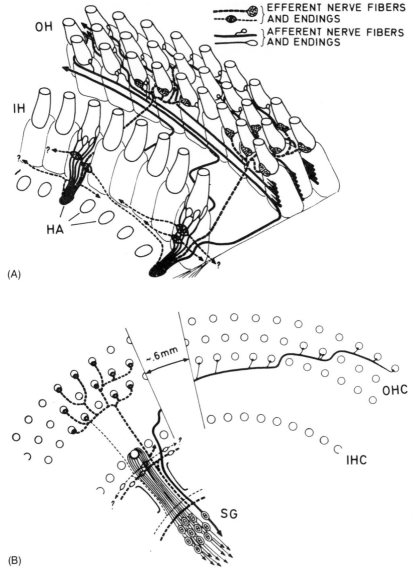

OH

IH

HA

(A)

EFFERENT NERVE FIBERS
AND ENDINGS
AFFERENT NERVE FIBERS
AND ENDINGS

~.6mm

OHC

IHC

SG

(B)

FIGURE 1.57. (A) Schema of the general innervation pattern of the organ of Corti. OH, outer hair cells; IH, inner hair cells; HA, habenular openings. (B) Schematic outline of the fiber distribution of the organ of Corti. Full thick lines: afferent fibers from the outer hair cells. Full thin lines: afferent fibers from the inner hair cells. Interrupted thick lines: efferent nerve fibers from the contralateral olivocochlear bundle. Interrupted thin lines: efferent nerve fibers from the contralateral olivocochlear bundle. SG, spiral ganglion. (From Spoendlin, 1970.)

and each outer hair cell is provided with approximately four afferent endings (Spoendlin, 1970, 1973). A similar arrangement seems to exist in the rat (Smith, 1973). The efferent terminals on the hair cells are also unevenly distributed (Figure 1.57B). There are more efferent terminals on each of the outer hair cells in the basal part of the cochlea than are on those in the apical turn of the cochlea. Each of the inner hair cells receives only a few efferent terminals. This vast number of efferent terminals is supplied by only 600–1000 efferent fibers of the olivocochlear bundle. Thus, each efferent fiber gives rise to many efferent terminals. There is even a third type of innervation to the sensory hair cells called adrenergic (autonomic) fibers. Information about this innervation is sparse.

Mechanical Properties of the Basilar Membrane

It was long assumed that the basilar membrane played an important role in the ability of the ear to discriminate tones of different frequency or sounds of different spectral composition. The basilar membrane is set into motion by vibrations of the surrounding fluid in the scala vestibuli and scala tympani. Theoretically, the resulting motion of the basilar membrane can be of several different types. On the basis of a small amount of experimental evidence, many hypotheses about the functioning of the cochlea were put forward. One of the early ones was Helmholtz' resonance theory (1863) in which he postulated that each narrow segment of the basilar membrane functioned as an individual *resonator*. A pure tone would thus give rise to vibration in only one segment. Later, this theory was modified to include a mechanical damping (Gray, 1900) of these resonators so that a pure tone would also produce vibrations in segments adjacent to the one having the highest vibration amplitude. However, it was not until it was possible to perform experiments in which the motion of the basilar membrane could directly be observed, that the wavelike nature of this motion was known (von Békésy, 1942). The motion of the basilar membrane is a result of an interaction between the basilar membrane and the surrounding fluid, and is governed by the mechanical properties of the basilar membrane, such as its stiffness, friction, and mass. Change in these constants along the membrane is particularly important in determining the motion of the membrane. In fact, as pointed out by von Békésy, different modes of motion of the membrane are possible, depending on how these constants vary along the membrane (von Békésy, 1955, 1956). Thus, the basilar membrane can function like a resonator in a way that the resonance frequency is different at different locations along the membrane, or, the vibration of the surrounding fluid can set up standing waves or traveling waves along the basilar membrane. There is, finally, the possibility that the basilar membrane has no spectral selectivity at all but vibrates in nearly the same way as a telephone diaphragm. This telephone theory assumes that frequency analysis is performed entirely in the central nervous system.

The *stiffness gradient* along the membrane is the parameter that has the greatest influence upon the pattern in which the basilar membrane actually vibrates. With exact knowledge about the values of the stiffness, mass, and friction at different locations along the membrane, it is possible to predict the type of vibration a sound will set up along the basilar membrane. Physical constants of the basilar membrane are, however, very difficult to determine because of several factors, including the small size of, and difficulty in exposing, the structure in question. These constants have not yet been measured with an accuracy sufficient to predict the motion of the basilar membrane.

It was a major breakthrough when von Békésy (1942), using a light microscope and stroboscopic light, observed the motion of the basilar membrane in human cadaver ears in response to pure tone stimulation. He found that the motion of the basilar membrane in response to a pure tone was a *traveling* wave which began at the base of the cochlea and continued toward the apex of the cochlea. Figure 1.58 shows a simplified physical model of the ear to illustrate the wave motion along the basilar membrane as we assume it occurs in response to a continuous tone (Zweig, Lipes, & Pierce, 1976). Proceeding from the base, the wave initially increased its amplitude as it traveled toward the apex. After a certain distance, its amplitude reached its maximum and then rapidly decreased (Figure 1.59). As the wave approached its maximum amplitude, its velocity decreased. The distance traveled by a given wave along the basilar membrane before it reached its maximum was found to be uniquely related to the frequency of the stimulus tone. Low-frequency tones travel a long distance before their amplitude decreases, whereas high-frequency tones travel only a short distance. It is thus possible to construct an entire frequency scale along the basilar membrane. If two tones of sufficiently different frequency are sounded simultaneously, then two amplitude maxima will be produced along the basilar membrane.

Hydrodynamic considerations tell us that the motion of the basilar membrane is governed by its own physical properties and by the interaction between the membrane and the surrounding fluid. In an ordinary

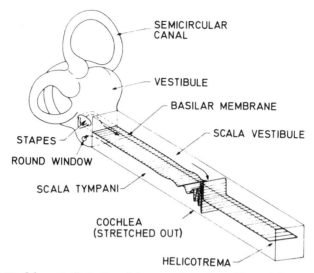

FIGURE 1.58. Schematic illustration of the ear to show the cochlea and illustrate the wave motion on the basilar membrane in response to a steady pure tone. (From Zweig *et al.*, 1976.)

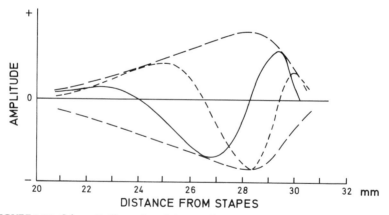

FIGURE 1.59. Schematic illustration of the traveling wave motion along the basilar membrane set up by a pure tone. (From von Békésy, 1947.)

mechanical resonance system, there is a cyclic exchange of energy between mass and stiffness. A similar exchange of energy between the membrane and the surrounding fluid occurs in the cochlea (Zwislocki, 1948). The traveling wave motion comes into being because stiffness and mass of the basilar membrane vary in a certain way along its length. Most important is the gradual decrease of the stiffness from base to apex. At the apex, stiffness of the basilar membrane is less than 1% of its value at the base.

On the basis of the observations just described, it may be assumed that the various spectral components of a complex sound give rise to displacement maxima along the basilar membrane, and hence to a vibratory pattern that corresponds to the spectral distribution of the energy in the sound. Information concerning the amount of energy in various freqency bands may, therefore, be transmitted to higher nervous centers by the way the hair cells along the basilar membrane are activated by the membrane motion. This method of coding auditory information may or may not have a dominant functional importance in the discrimination of sounds such as speech sounds. This matter will be discussed in further detail in Chapters 3 and 4.

FREQUENCY SELECTIVITY OF THE BASILAR MEMBRANE

Measurements by von Békésy (1942) showed that at a certain point along the basilar membrane the vibration amplitude has its maximal value at one specific stimulus frequency, and declines gradually when the frequency of the stimulus tone is increased or decreased. Frequency width of such frequency response functions determines the frequency resolution of the basilar membrane as a spectrum analyzer, as for example in the ability of the basilar membrane to separate the individual spectral components of a complex sound. Response amplitudes at different points along the membrane are shown in Figure 1.60.

When von Békésy's data were first published in 1942, it seemed clear that the frequency selectivity of the basilar membrane was not sharp enough to explain the actually existing psychoacoustical frequency selec-

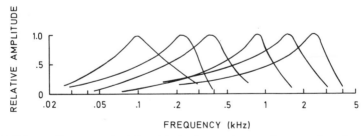

FIGURE 1.60. Frequency selectivity of the basilar membrane as it appears from the studies of von Békésy. Each curve represents the relative vibration amplitude at various points along the membrane in a human cadaver ear. (From von Békésy, 1942.)

tivity. It seemed natural to seek an explanation in the nervous system, where, it was suggested, an additional sharpening of the frequency resolution occurred. Results obtained in studies of other sensory systems suggested that "lateral inhibition" should also be present in the ear, and this process would be likely to sharpen the mechanical tuning so to account for all the observations made up to that time. Since then, many hypotheses have been put forward to explain the mechanism behind the great frequency resolution of the auditory system. These questions are far from resolved, but the results of recent studies in which new techniques, such as recording of potentials from the interior of hair cells have been possible, have shed new light on the problems. The new results are dealt with in Chapter 3.

It is argued that von Békésy's results did not reflect the function of the basilar membrane at physiological sound intensities. In order to achieve observable amplitudes, von Békésy had to set the cochlear fluid into a motion—the amplitude of which corresponded to a sound level of about 145 dB SPL—which is far higher than that handled by the ear under physiological conditions. This limit was imposed by the wavelength of the stroboscopic light used to illuminate the basilar membrane. It is now possible to measure the vibration amplitude of the basilar membrane at much lower and at more physiological sound intensities. One of the methods for such measurements is the use of the Mössbauer effect. This method allows the measurement of smaller amplitudes than does light microscopy with the limiting factor being that the wavelength of the light is approximately 3000 Å. Using the Mössbauer method, it is possible to measure velocities as small as .2 mm/sec. For sinusoidal motion at 1000 Hz, this corresponds to an amplitude of approximately 200 Å, and at 10 kHz, of about 20 Å. This method was first used to study the vibration of the basilar membrane in the guinea pig by Johnstone and Boyle (1967). Figure 1.61 shows how the Mössbauer method was applied to the measurement of the vibration of the basilar membrane of the squirrel monkey by Rhode (1971).

Results obtained by Rhode (1971) using the Mössbauer method in the anesthetized squirrel monkey showed a narrower tuning, (i.e., greater selectivity) than the curves presented by von Békésy (1942). Rhode could show that the vibration of the basilar membrane was nonlinear in such a way that the tuning curves were narrower at low intensities than at high. Frequency transfer functions, showing the ratio between the vibration amplitude of the basilar membrane and that of the malleus obtained by Rhode at 70, 80, and 90 dB SPL, respectively, are shown in Figure 1.62. It

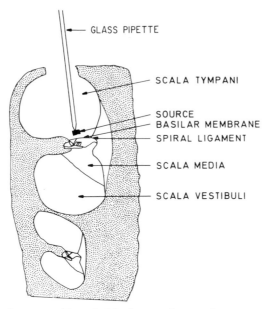

GLASS PIPETTE

SCALA TYMPANI

SOURCE
BASILAR MEMBRANE
SPIRAL LIGAMENT

SCALA MEDIA

SCALA VESTIBULI

FIGURE 1.61. Illustration of how the Mössbauer technique of measuring vibration is applied to measurements of the motion of the basilar membrane in the ear of a squirrel monkey. The radioactive source used was 40 × 60μm. (From Rhode, 1971.)

is clearly seen that these curves do not have the same shape, indicating that the motion of the basilar membrane is nonlinear. (The curve obtained at 80 dB is shifted downward 10 dB, and that obtained at 90 dB is shifted 20 dB downward, to facilitate comparison.) The three curves would have coincided in a linear system. Further studies by Rhode (1973) showed that this nonlinearity is "physiologically vulnerable," which means that the nonlinearity disappears as a result of anoxia.

Rhode (1973) finds that postmortem resonance of the basilar membrane is broader than it is in the live preparation and that the frequency of the peak response of a certain point shifts downward in frequency. The larger amplitude at low frequencies postmortem was taken as an indication that the compliance increased after death (Zwislocki, 1980). It has not been possible yet to describe the in vivo response mathematically (Zweig et al., 1976), whereas it is possible for the postmortem response.

Direct comparison of the results of von Békésy (1942) with those of Rhode (1971) cannot be made because the former measurements were made on the low-frequency parts (50–2400 Hz) of the basilar membrane,

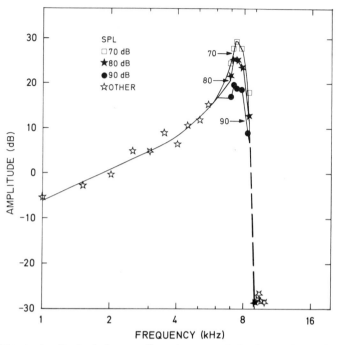

FIGURE 1.62. Amplitude of vibration of a single point on the basilar membrane of a squirrel monkey measured by the Mössbauer technique. The curves show the amplitude of vibration for constant displacement amplitude of the malleus. These curves are usually named transfer functions (transfer ratios). The three curves represent three different sound intensities. Two of the curves are shifted vertically in such a way that all three curves should coincide if the system had been linear. (From Rhode, 1971.)

whereas the latter measurements have made on the high-frequency parts (near 8 kHz).

The nonlinearity shown by Rhode (Rhode, 1971; Rhode & Robles, 1974) has attracted attention for several reasons. It might explain on the basis of other results regarding the frequency selectivity of the auditory periphery why the tuning curves in human cadaver ears obtained by von Békésy (1942) were much broader than expected. This nonlinearity might also explain the distortion observed in both psychoacoustic and neurophysiological experiments when two tones were presented simultaneously. It is particularly intriguing that other studies using the Mossbauer effect or other techniques have been unable to find a nonlinear function of the basilar membrane (Johnstone & Taylor, 1970; Johnstone, Taylor, & Boyle, 1970; Wilson & Johnstone, 1972, 1975.)

Results obtained using the capacitance probe for measuring basilar membrane displacement in anesthetized animals indicate that the basilar membrane vibration is generally linear (Wilson & Johnstone, 1972, 1975). Contrary to results of von Békésy (1942) and Rhode (1971), which show a frequency tuning, the studies just mentioned show that the basilar membrane essentially functions as a lowpass filter without any pronounced tuning. An average transfer function of a single point on the basilar membrane obtained in the guinea pig using this method is shown in Figure 1.63.

There is thus considerable discrepancy between the results of measurement of the vibration of the basilar membrane. Although some of the recent studies (Rhode, 1971) show that the basilar membrane has a higher degree of selectivity than earlier results (von Békésy, 1942) indicated, it is generally recognized that the membrane's mechanical frequency selectivity is not sufficient to explain the entire spectral selectivity of the auditory system as it emerges from other studies such as psychoacoustic studies

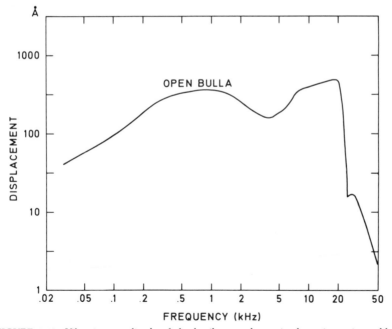

FIGURE 1.63. Vibration amplitude of the basilar membrane in the guinea pig cochlea measured by the capacitive probe technique. This curve shows the average vibration amplitude of the basilar membrane corrected for constant sound pressure level (100 dB SPL) at the tympanic membrane with the bulla open. (From Wilson & Johnstone, 1975.)

and neurophysiological studies, for instance recordings from single auditory nerve fibers. It may be mentioned that the response of the basilar membrane to a short impulse also shows nonlinearities when measured using the Mössbauer effect (Figure 1.64). We will consider frequency selectivity and nonlinearity of the basilar membrane in further detail in Chapter 3.

It may suffice to point out in this section that, in discussing frequency selectivity of the basilar membrane, it should be kept in mind that all measurements of the basilar membrane motion refer to the up-and-down vibration amplitude of one single point at a time. It is not obvious that it is this simple up-and-down movement of the basilar membrane that constitutes the efficient stimulation of the hair cells. The vibration pattern of the basilar membrane is very complex, and it has been suggested that such attributes of the basilar membrane motion as *shear* waves or radial displacements may be efficient stimuli to hair cells (von Békésy, 1953a, b; Duifhuis, 1976). How these wave motions are related to the up-and-down displacement of a single point is not completely understood, and it is possible that the excitation of hair cells depends on parameters other than the up-and-down motion of a single point.

VIBRATORY AMPLITUDE OF THE BASILAR MEMBRANE AT THRESHOLD

At physiological sound levels, the absolute value of the vibratory amplitude of the basilar membrane is extremely small. On the basis of von Békésy's results (1948, 1960, p. 173) using the light microscope technique, threshold displacement is estimated to be .003 Å for a point on the basilar membrane that has its greatest amplitude at 3 kHz for a pure tone with that frequency. These values were obtained by linear extrapolation from measurements of vibratory amplitudes for a sound intensity of more than 140 dB SPL and are based on the assumption that the basilar membrane vibrates in a linear way. However, as this assumption is not valid, these values of vibration amplitude at threshold obtained by such extrapolation are probably far from being correct.

Results obtained using the Mössbauer technique in the squirrel monkey show that the basilar membrane has a vibration amplitude of 1000–3000 Å at the point where 7 kHz gives maximal vibration amplitude for a 7 kHz tone at a sound level of 100 dB. The behavioral threshold at that frequency is about 20 dB SPL. Similar values were obtained in the guinea pig (Rhode, 1978). If a linear extrapolation is valid, the root mean square vibration amplitude at the threshold of hearing would amount to about .1–.3 Å at the frequency (7 kHz) and the peak-to-peak displacement

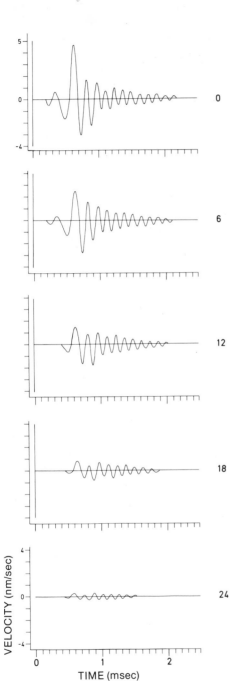

FIGURE 1.64. Impulse response (velocity in nanometers per second) of the basilar membrane vibration at different sound intensities (given in relative decibels by legend numbers). The results were obtained from an anesthetized squirrel monkey using the Mössbauer technique and click stimulation. (From Robles *et al.*, 1976.)

would be about three times as great. Results of Wilson and Johnstone's work (1975), also using linear extrapolation from the results of measurements at high sound levels, indicate that the threshold displacement is about 3 Å at 20 kHz in the guinea pig. The behavioral threshold for the guinea pig at 20 kHz is about 50 dB SPL. These values are larger than those measured by von Békésy (1948) in human cadaver ears. If the nonlinearity shown by Rhode (1971, 1973) is accepted, rather different vibratory amplitude values are obtained at threshold intensities.

Sensory Transduction in Hair Cells

Mechanical exitation of hair cells is assumed to control the discharge pattern in the afferent (ascending) nerve fibers forming the auditory part of the eighth cranial nerve. The exact mechanism of the transformation of mechanical motion into excitation of the hair cells is not known, but the process can be thought of as a sequence of three main steps. The first step is presumably a mechanical deflection of the hairs that results in a change in the intracellular potential of the hair cell (receptor potential). The next step invovles the chemical transmission at the synapse that connects the hair cell to the afferent nerve fiber. The third step consists of setting up discharges in the afferent nerve fiber, the rates of which are controlled by the generator potential that appears in the dendritic region.

Many theories have been put forward to describe the transduction of mechanical energy into neural excitation in the hair cells. Davis (1965) suggested that the deflection of the hairs of the hair cells resulted in a resistance change that, in turn, changed the current flowing through the hair cells. The primary source of this current was assumed to be the "battery" represented by the endocochlear potential (see Chapter 2).

The most reliable way of obtaining information on the excitation in the hair cells would seem to be through recording the intracellular potentials in the hair cells. Hair cells of the mammalian cochlea are, however, very small and such recordings have been made possible only recently (Russell & Sellick, 1978). (See p. 91.)

Much of the knowledge about the function of hair cells has been gained through studying hair cells in the lateral line organ of fish because these are larger and more easily accessible than those in the mammalian cochlea (Flock 1965a, b, 1971). Results of such experiments are schematically illustrated in Figure 1.65. Observations made in the lateral line hair cells of fish are assumed to be relevant for the hair cells in the mammalian cochlea

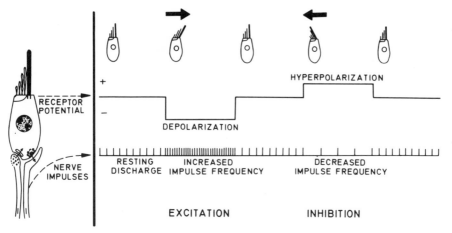

FIGURE 1.65. Schematic illustration of the excitation of hair cells. Deflection in one direction causes depolarization of the interior of the cell and subsequent increase in the discharge rate (excitation), whereas deflection in the opposite direction causes a hyperpolarization followed by a decrease in discharge rate (inhibition). (From Flock, 1965b.)

because of the similarities between the hair cells in the lateral line organ and those in the cochlea.

Studies of hair cells of the lateral line organ seem to indicate that these cells transduce deflection of the hairs as a half-wave rectifier. In other words, they exhibit excitation in only one direction. Movements in the opposite direction decrease the discharge rate of afferent nerve fibers and are thus regarded as inhibitory (Flock, 1965b). In the lateral line organ of the fish, hair cells are excited when the hairs are displaced in the direction of the basal body (see Figure 1.66). Deflection in that direction accelerates the discharge rate of the nerve fiber leading from the hair cell, whereas the sensitivity decreases as the direction of the displacement is shifted away from that orientation. Recordings from single nerve fibers of the auditory nerve in mammals indicate that mammalian hair cells act similarly. However, studies of intracellular potentials of inner hair cells in the guinea pig indicate that the transduction of sound is not precisely a half-wave rectification, but has the form of a nonlinearity, with the slope for motion in one direction of the input–output functions being approximately one third of the slope of the input–output function, that corresponds to motion in the opposite direction (Sellick & Russell, 1980; Russell & Sellick, 1981).

Despite efforts to describe the chemical substance that transmits infor-

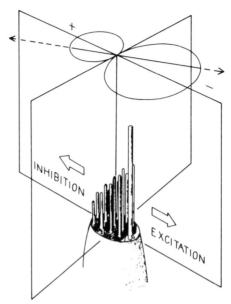

FIGURE 1.66. Directional sensitivity of a hair cell. (From Flock, 1965a.)

mation over the synaptic cleft in the afferent synapse (transmitter substance), still not much is known about it. Glutamate and aspartate (Godfrey, Carter, Berger, & Matchinsky, 1976; Klinke & Oertell, 1977) have been suggested as likely substances to fulfill this function, as have the catecholamines (Thornhill, 1974) and gamma amino butyric acid—GABA (Flock & Lam, 1974). However, Comis and Leng (1979) found evidence that glutamate could fulfill this role, and they ruled out adrenaline as the afferent transmitter substance in the hair cells.

It is not known in detail how the motion of the basilar membrane is transferred to the hairs of hair cells. Several hypotheses have suggested that the tectorial membrane plays a part in transferring the movement of the basilar membrane to hair cells. One hypothesis assumes that the tectorial membrane slides over the reticular lamina when the basilar membrane vibrates. Because the hairs of the outer hair cells are embedded in the tectorial membrane, they are then deflected. Another hypothesis states that when the tectorial membrane slides over the reticular lamina it causes the perilymph to flow, which then sets the hairs of hair cells into motion (Billone, 1972).

Electron microscopic work by Lim (1972) has shown that the stereocilia of outer hair cells are firmly embedded in the tectorial membrane, whereas those of the inner hair cells are not. This difference in the coupling between the tectorial membrane and the hairs of inner and outer hair cells implies that two populations of hair cells must have different responses to basilar membrane motion. Dallos (1973) found that, when some of the outer hair cells were destroyed by the ototoxic antibiotic kanamycin, the cochlear microphonic potentials in regions of the cochlea where the outer hair cells were destroyed were related to the velocity of the basilar membrane motion. The cochlear microphonic potential in regions with intact outer hair cells reflected the basilar membrane displacement. This led Dallos and Cheatham (1976) and Dallos, Billone, Durrant, Wang, and Raynor (1972) to propose that the inner hair cells respond to basilar membrane velocity and the outer hair cells respond to basilar membrane displacement.

When Sellick and Russell (1980) recorded intracellular potentials from inner hair cells, they found that inner hair cells responded to basilar membrane velocity for low frequencies (below 100–200 Hz) but responded to basilar membrane displacement for high frequencies. They considered the lowpass function of the hair cell—the equivalent time constant of which was found to be about .22 msec. One of the consequences of the fact that the inner hair cells respond to the velocity of the basilar membrane for low frequencies will be that the hair cells are protected from large displacements by slow and large deflections of the basilar membrane.

It has been proposed by Davis (1957) that the effective stimulus for frequencies about 200 Hz is the shear displacement of the stereocilia. This is in agreement with the results of experiments conducted in the sacculus of the frog by Flock (1971).

It has been shown that the hairs contain actin, which implies that their stiffness may somehow be controlled (Flock & Cheng, 1977; Flock, 1980). How this may contribute to neural transduction in the cochlea is unknown.

It was assumed that movement of the basilar membrane toward the scala vestibuli was excitatory. This assumption was based on the observation that the latency of the neural response recorded from the round window compound action potential (CAP) was shortest in response to a rarefaction click. A rarefaction click causes an outward movement of the stapes, thus a movement of the basilar membrane toward the scala vestibuli (Kiang & Peake, 1960). Results of studies using trapezoidal waveforms as stimuli indicate that the excitatory phase is represented by a movement of the basilar membrane in the opposite direction—toward the scala tympani (Zwislocki, 1974; Konishi & Nielsen, 1978). These studies also indicate that inner and outer hair cells may be excited by deflection of the basilar membrane in different directions (Zwislocki & Sokolich, 1974).

In the discussion of the anatomy of hair cells it was mentioned that these cells evolved from the lateral line organ in fish, where they first were surface receptors, but soon became located in canals to which the outside fluid had only indirect access. From that point in evolution, the fluid surrounding what became mammalian cochlear hair cells, has been very similar to endolymph—rich in potassium and poor in sodium. It seems essential, then that the hairs be surrounded by such fluid, although exactly what this has to do with the function of the hair cells is unknown.

There is an efferent (descending) neural input to the hair cells that originates from nerve fibers traveling together with the afferent fiber. This neural input is known as the Rasmussen bundle, and stimulation of this bundle is generally assumed to decrease the sensitivity of the hair cells. In the cat, electrical stimulation of the efferent nerve fibers has been shown to reduce the discharge rate of auditory nerve fibers in response to tone stimuli (Fex, 1962; Wiederhold, 1970; Wiederhold & Kiang, 1970). This inhibition occurs with a relatively long latency and is probably accomplished through the release of a specific transmitter substance (Fex, 1973; Klinke & Galley, 1974).

Electrical Potentials in the Cochlea

Two types of electrical potentials have been demonstrated in the cochlea: (a) dc potentials that are independent of the sound stimulus; and (b) ac potentials, and slowly varying dc potentials that are related to the sound stimulus.

POTENTIALS INDEPENDENT OF SOUND STIMULATION

As an electrode that is suitable for the recording of dc potentials is advanced through the different structures of the cochlea in the manner illustrated in Figure 1.67, various shifts in dc potentials are noted. There is a potential (endocochlear potential or EP) of about $+50$ mV between the inside of the scala media and the scala tympani, and within the organ of

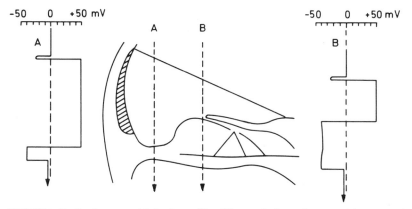

FIGURE 1.67. Resting potentials in the cochlea. The graph shows the potentials measured by an electrode as it travels from the scala vestibuli through Reissner's membrane to the scala media and through the organ of Corti towards the scala tympani. Two different electrode tracks are shown. (From von Békésy, 1952.)

Corti there is a negative potential of approximately 50 mV. The respective potentials of the scala vestibuli and the scala tympani are nearly the same (von Békésy, 1952).

The endocochlear potential is assumed to originate in the *stria vascularis* (see Figure 1.53) and is not a result of the particular ionic composition of the cochlear fluid. The presence of this potential seems to be essential to the normal functioning of hair cells.

POTENTIALS EVOKED BY SOUND

Cochlear Microphonics (CM)

By far, the most studied sound dependent potential is the cochlear microphonic (CM) potential. It was first recorded in the internal auditory meatus—in the auditory nerve (Wever & Bray, 1936), but it can also be recorded at various other places in the inner ear, such as near the round window and in the cochlea itself.

As the name indicates, cochlear microphonic is a potential similar to that produced by a microphone—it reproduces the waveform of a sound. An electrode connected to the round window of an experimental animal and connected to a suitable amplifier and loudspeaker will reproduce the voice of a speaker that talks into the ear of the animal. Since the discovery of this potential, numerous studies have been devoted to its characteristics. The cochlear microphonic potential has been recorded mainly in two ways—by a gross electrode near or on the round window, or by intracochlear electrodes, usually bipolar. In either case, the response to a pure tone is mostly a sine wave with little distortion present. The amplitude of the sine wave increases in proportion to increased sound pressure level at the tympanic membrane up to a certain saturation point beyond which the CM no longer increases (Figure 1.68). The cochlear microphonic potential has no apparent threshold. The lowest CM level recordable seems to be limited only by the noise in the recording apparatus and the bandwidth of the analysis apparatus.

There are most likely other secondary sources of cochlear microphonics which, under normal conditions and compared to the main source of microphonics, contribute too little to be discerned. However, when an animal is subjected to asphyxia, these weaker cochlear microphonics can be measured. After death, the CM from secondary sources can be measured for as long as several hours.

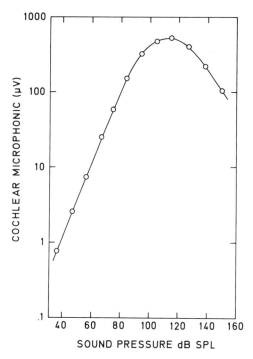

FIGURE 1.68. The amplitude of the cochlear microphonic potential in response to a pure tone recorded near the round window in a cat as a function of the sound intensity. (From Dallos, 1973.)

The origin of the cochlear microphonic potential is most likely the hair cells. Studies indicate that the CM potential recorded near the round window is likely to originate mainly in the outer hair cells in the basal part of the basilar membrane (Dallos, 1973, p. 276). The amplitudes of the CM potentials have been shown to approximate the velocity of the cochlear fluid (Møller, 1963). This makes the cochlear microphonic potential a valuable tool in studying other functions of the ear in experimental animals.

The question of whether the CM is related to displacement of the basilar membrane, velocity, or acceleration of the hair cells has been disputed. Results obtained by von Békésy (1951) indicate that a radial displacement of the hairs in more efficient in producing CM than other directions of motion. Since the CM measured in the manner of von Békésy is dominated by the response from outer hair cells, it may be concluded that radial displacement is the most efficient stimulus for outer hair cells. Some results indicate that the inner hair cells generate CM in proportion to the *velocity* of displacement of the basilar membrane (Dallos, 1973, p. 274).

The results of recording from the interior of inner hair cells have shed new light on the generation of cochlear microphonics and the transduction process in the hair cells. Sellick and Russell (1980) compared the intracellular receptor potentials measured in inner hair cells with the cochlear microphonics (CM) recorded from the scala tympani and the sound pressure at the tympanic membrane in guinea pigs. They found that, in response to low-frequency tones (100 Hz), the intracellular potential is a distorted sine wave with a depolarization phase exceeding the hyperpolarization phase by a factor of three. They concluded that inner hair cells respond to the basilar membrane velocity for frequencies below 100–200 Hz. Above that frequency, the effective stimulus for inner hair cell response is basilar membrane displacement. Sellick and Russell's study also showed that the CM that can be recorded in the scala tympani (and at the round window) represents predominantly the response of outer hair cells. These results are in agreement with findings by Dallos (1973, p. 276), who showed that the potentials generated by displacing the inner hair cells are smaller than those generated by displacing the outer hair cells. Therefore, the CM generated by the inner cells may not be distinguished in the normal cochlea.

Intracochlear recordings using bipolar electrodes make it possible to record from hair cells located within limited sections of the basilar membrane. Many such experiments have been performed, usually in the guinea pig, since the cochlea of the guinea pig protrudes from the wall of the middle ear, making it easily accessible (see Dallos, 1973). The location on the basilar membrane from which recordings are made is determined by the location of the electrodes. Recorded in this way, the CM display frequency selectivity properties that have their origin in the mechanical frequency selectivity of the basilar membrane, but are not necessarily identical to it.

Using a variation of this technique, different degrees of frequency selectivity may be seen in the recorded potentials. Sharpness of the tuning is clearly related to the exact technique used. The prevailing method consists of placing one electrode in the scala tympani and one in the scala vestibuli—one on each side of the basilar membrane. These electrodes are then connected to a differential amplifier which records the difference in potentials between the two electrodes. Recordings of the cochlear microphonic potential using this method may be affected by cochlear microphonics generated in the adjacent turns of the cochlea. Therefore, other methods such as a micropipette placed in the scala media (Zwislocki, 1980) or two electrodes placed close to each other in the same scala have been used (Pierson & Møller, 1980a, b).

Results of experiments (Pierson & Møller, 1980a, b) indicate that the CM recorded intracochlearly may originate from two different types of generators that may be inner and outer hair cells, respectively. One of these sources was found to be more sensitive to anoxia and fatigue than the other. The two sources were 180° out of phase, and in the normal cochlea the interaction between the two sources manifested itself by a sharp dip in the curve, showing the amplitude of the CM as a function of intensity. In these experiments, the CM was measured in a narrow frequency range one-fifth to one-half octave above the frequency of maximal response (Pierson & Møller, 1980a). Fatigue or hypoxia could eliminate the low-intensity source of CM. Using a low-frequency tone, together with a high-frequency one as stimuli, it is possible to study the effect on the CM that results from the high-frequency tone from the displacement of the basilar membrane caused by the low-frequency tone. Such experiments revealed in the unfatigued cochlea that the CM was enhanced when the basilar membrane was deflected toward the scala tympani, while in the fatigued cochlea the opposite was the case, and the CM was enhanced when the basilar membrane was deflected toward the scala vestibuli (Pierson & Møller, 1980b). These experiments thus suggest that there are two populations of generators of cochlear microphonics, and that they may be inner and outer hair cells.

Since 95% of all afferent auditory nerve fibers innervate inner hair cells, the role of the outer hair cells has been hypothetized to be that of modulating—either by enhancing or inhibiting the function of the inner hair cells. It is thus possible that the modulation of the CM caused by deflecting the basilar membrane, as in the previously mentioned experiments, is a result of such interaction. Such a relationship between inner and outer hair cells, in which they both oppose each other and interact with one another, has been suggested by several investigators (Zwislocki & Sokolich, 1974; Eldredge, 1976; Sokolich, Hamernik, Zwislocki, & Schmiedt, 1976).

Summating Potentials (SP)

In addition to the cochlear microphonic potential, a dc potential may be recorded in response to a continuous tone (Davis, Fernandez, & McAuliffe, 1950; von Békésy, 1950). When a tone burst is presented, this dc potential usually appears as a shift in the baseline of the CM (Figure 1.69). This dc shift (summating potential or SP) can be measured at various places in and on the cochlea, but it is most prominent when recorded in the scala media. The summating potential is not nearly as closely correlated to the stimulus as is the CM. Thus, the SP is a variable quantity, and it may even have different polarities in seemingly similar experimental situations (see Dallos, 1973).

Several theories about the origin of the SP have been put forward. The most credible one to date is that the SP is a result of a distortion caused by an asymmetry in the mechansims producing CM. This hypothesis regards summating potentials as distortion products, and it is suggested that they originate when displacement of the hair cells in one direction produces more electrical current than displacement in the opposite direction (Whit-

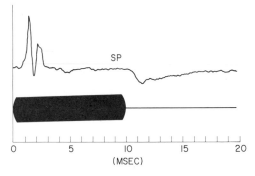

FIGURE 1.69. Response from the round window of a guinea pig to a 20-kHz tone burst, showing the summating potential.

field & Ross, 1965). Other hypotheses claim that the SP consists of several components, originating from different sources such as inner hair cells (negative SP), and outer hair cells (positive SP). These two sources may contribute different proportions of the SP under different stimulus conditions. This would explain the variable character of this potential and would particularly explain the effect of anoxia on the potential (Dallos, 1973, p. 290).

Compound Action Potentials (CAP)

Another sound-dependent potential is the compound action potential (CAP) which reflects neural activity in the auditory nerve. It may be recorded both in the cochlea and its vicinity, and from the auditory nerve, the latter being technically more difficult. This potential is believed to reflect not only the degree of excitation but also, to a large extent, the degree of synchronization of the activity of the auditory nerve fibers. This potential is therefore most prominent in response to transient sounds such as click sounds and high-frequency tone bursts with rapid rise times. Figure 1.70 shows the CAP response recorded at the round window of a rat. As can be seen, it has two pronounced negative peaks which are usually named N_1 and N_2. Some investigators assume that N_2 originates in the cochlear nucleus and that it is the mass response from secondary neurons, while others assume that it is a part of the primary response. The amplitude of N_1 is a function of both the stimulus intensity and the number of fibers firing in synchrony. Its amplitude is largest when many

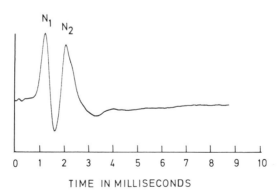

FIGURE 1.70. Compound action potential in response to a click sound recorded from the round window of a rat.

fibers are firing in synchrony, as is the case in response to sound with a steep rise, as, for instance, click sounds. Both amplitude and latency of the response vary in a characteristic way with the intensity of a sound. Figure 1.71 shows the latencies of the N_1 and N_2 peaks in a rat. The stimulus was bandpass-filtered clicks and the different curves refer to different center frequencies of the filter. There is a decrease in latency with increased intensity and latency is a function of the center frequency of the bandpass-filtered click sounds. Thus, the response to a low-frequency click has a much longer latency, and there is a larger decrease in latency with intensity than what is seen in the response to a highpass-filtered click.

Figure 1.72 shows the amplitudes of the N_1 and N_2 peaks as functions of intensity to two different bandpass-filtered clicks (center frequencies 20 kHz and 6.3 kHz). The amplitude increases monotonically with sound intensity until it reaches a plateau at high intensities. Irregularities seen in earlier recordings of the round window compound action potentials (Peake & Kiang, 1962) may be ascribed to the acoustical properties of the

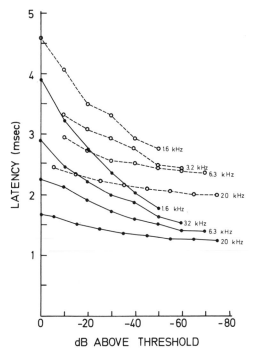

FIGURE 1.71. Latency of the N_1 (solid lines) and N_2 (dashed lines) peaks of the response to bandpass-filtered click sounds. The results were obtained from a rat and the filter was a 1/3-octave bandpass filter. The center frequency of the filter is indicated by legend numbers. The duration of the clicks before filtering was 5 μsec.

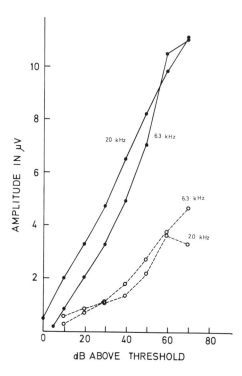

FIGURE 1.72. Amplitude of the N_1 and N_2 peaks in response to bandpass-filtered clicks as a function of stimulus intensity for two center frequencies (6.3 and 20 kHz). The results are from the same experiment as the data shown in Figure 1.71.

stimuli used and to the irregular frequency characteristics of the sound source used in earlier experiments.

Since the CAP originates in the auditory nerve, it possesses the frequency selectivity of the ear. Also, unlike the CM, the CAP has a rather well-defined threshold. Using modern averaging techniques, it is thus possible to determine this threshold as a function of stimulus frequency when tone bursts or bandpass-filtered clicks are used as stimuli. Using that technique, it has been shown that the threshold of the CAP recorded from the round window, evoked by tone bursts or bandpass-filtered clicks, is nearly parallel to the behavioral threshold of the animal (Dallos, Harris, Özdamar & Ryan, 1978); that is shown in Figure 1.73, which describes the CAP threshold and the behavioral threshold in a guinea pig. Also shown is the sound intensity required to produce a certain amplitude of the CM. The relatively good correlation between the results of electrophysiological methods and those of behavioral techniques in measuring hearing threshold has made CAP an important measure for determin-

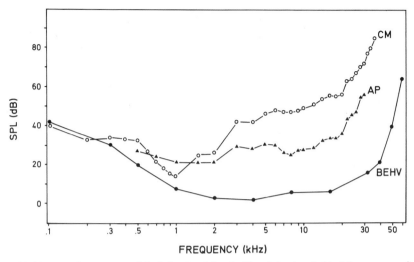

FIGURE 1.73. Comparison of the behavioral threshold and the threshold of the compound action potential (AP) for pure tones in a guinea pig. The sound pressure required to produce a certain amplitude of the CM is also shown. (From Dallos *et al.*, 1978.)

ing changes in the hearing thresholds of animals as a result of noise exposure, ototoxic drugs, etc. This is because the CAP threshold is easier to measure than the behavioral threshold.

There is a difference between the CAP measure of threshold and the behavioral method that may be worth noting since it does not directly follow from data such as those illustrated in Figure 1.73. Since the CAP threshold represents the function of only the peripheral part of the auditory system, it is influenced only by the temporal integrative properties of the periphery. Whereas the behavioral threshold involves the entire auditory system, and consequently is influenced by the temporal integration that occurs in the entire system. Temporal integration time of the peripheral auditory system is much shorter than the behavioral integration time. The former is on the order of one to a few milliseconds, whereas the time to integrate loudness near threshold is about 200 msec for the behavioral response. This variance in integration time may explain the difference in threshold as measured by behavioral techniques and that measured by electrophysiological techniques. Integration of loudness measured behaviorally is about 100 times greater than that measured in the peripheral auditory system. This corresponds to a 20-dB intensity difference; this is very close to the discrepancy between the CAP threshold and the behavioral threshold for frequencies above 2 kHz (see Figure 1.73).

All three sound-dependent potentials (CM, CAP, and SP) may be recorded from various locations on and in the cochlea. The graphs in Figures 1.68, 1.69, and 1.70 show these three potentials separately. It is natural to expect that each record show all three potentials simultaneously, but the recording of each of the three potentials is accomplished by making a suitable choice of sound stimulus in each case, and these stimuli are different. The CM is best recorded in response to a continuous tone while the CAP is most clearly seen in response to a transient sound. SP is also seen in response to a continuous pure tone but, because most recordings are done with ac coupled amplifiers, the summating potential is usually not seen when continuous tones are used for the purpose of recording the cochlear microphonic potential. Transient sounds also evoke an SP but it is so short that it is not always seen in records of CAP. Figure 1.74 shows a recording in which all three potentials can be seen. The stimulus is a tone burst of a relatively high intensity, and the recording amplifier had a long time constant to allow for recording of the summating potential.

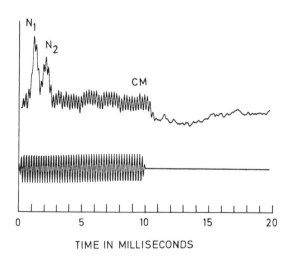

TIME IN MILLISECONDS

FIGURE 1.74. Recording from the round window of a rat to show CM, CAP (N_1, N_2), and SP at the same time. The stimulus (10-msec long) was a 5-kHz tone burst.

References

Ballentyne, S. Effect of diffraction around the microphone in sound measurements. *Physiological Reviews*, 1928, *32*, 988–992.

Békésy, G. von Zur Physik des Mittelohres und über das Hören bei fehlerhaften Trommelfell. *Akustische Zeitschrift*, 1936, *1*, 13–23.

Békésy, G. von Über die Messung der Schwingungsamplitude der Gehörknochelchen mittels einer kapacitiven Sonde. *Akustische Zeitschrift*, 1941, *6*, 1–16.

Békésy, G. von Über die Schwingungen der Schneckentrennwand beim Präparat und Ohrenmodell. *Akustische Zeitschrift*, 1942, *7*, 173–186.

Békésy, G. von The variation of phase along the basilar membrane with sinusoidal vibrations. *Journal of the Acoustical Society of America*, 1947, *19*, 452–460.

Békésy, G. von Vibration of the head in a sound field, and its role in hearing by bone conduction. *Journal of the Acoustical Society of America*, 1948, *20*, 749–760.

Békésy, G. von The vibration of the cochlear partition in anatomical preparations and in models of the inner ear. *Journal of the Acoustical Society of America*, 1949, *21*, 233–245.

Békésy, G. von DC potentials and energy balance of the cochlear partition. *Journal of the Acoustical Society of America*, 1950, *22*, 576–582.

Békésy, G. von Microphonics produced by touching the cochlear partition with a vibrating electrode. *Journal of the Acoustical Society of America*, 1951, *23*, 29–35.

Békésy, G. von DC resting potentials inside the cochlear partition. *Journal of the Acoustical Society of America*, 1952, *24*, 72–76.

Békésy, G. von Description of some mechanical properties of the organ of Corti. *Journal of the Acoustical Society of America* 1953. *25*, 770. a

Békésy, G. von Shearing microphonics produced by vibrations near the inner and outer hair cells. *Journal of the Acoustical Society of America*, 1953. *25*, 786. b

Békésy, G. von Paradoxical direction of wave travel along the cochlear partition. *Journal of the Acoustical Society of America*, 1955. *27*, 155–161.

Békésy, G. von Simplified model to demonstrate the energy flow and formation of traveling waves similar to those found in the cochlea. *Proceedings of the National Academy of Sciences USA*, 1956, *42*, 930–944.

Békésy, G. von *Experiments in hearing.* New York: McGraw-Hill, 1960.

Billone, M. C. *Mechanical stimulation of cochlear hair cells.* Unpublished thesis, Northwestern University, Evanston, Ill., 1972.

Borg, E. A quantitative study of the effect of the acoustic stapedius reflex on sound transmission through the middle ear of man. *Acta Otolaryngologica* (Stockholm) 1968, *66*, 461–472.

Borg, E. On the nonlinear dynamic properties of the acoustic middle ear reflex of unanesthetized animals. *Brain Research*, 1971, *31*, 211–215.

Borg, E. On the neuronal organization of the acoustic middle ear reflex, a physiological and anatomical study. *Brain Research*, 1973, *49*, 101–123. a

Borg, E. Neuroanatomy of brainstem auditory system of the rabbit. *Brain Research*, 1973, *52*, 424–427. b

Borg, E., & Møller, A. R. Effect of ethyl alcohol and pentobarbital sodium on the acoustic middle ear reflex in man. *Acta Otolaryngologica* (Stockholm), 1967, *64*, 415–426.

100 1. ANATOMY AND GENERAL PHYSIOLOGY OF THE EAR

Borg, E., & Møller, A. R. Effect of central depressants on the acoustic middle ear reflex in rabbits. *Acta Physiologica Scandinavica* (Stockholm), 1975, *94*, 327–338.

Borg, E., & Zakrisson, J. E. Stapedius reflex and speech features. *Journal of the Acoustical Society of America*, 1973, *54*, 525–527.

Borg, E., & Zakrisson, J. E. The activity of the stapedius muscle in man during vocalization. *Acta Otolaryngologica* (Stockholm), 1975, *79*, 325–333.

Bosher, S. K. The nature of the negative endocochlear potentials produced by anoxia and ethacrynic acid in the rat and guinea-pig. *Journal of Physiology* (London), 1979, *293*, 329–345.

Brodel, M. *Three unpublished drawings of the anatomy of the human ear*. Philadelphia: W. B. Saunders, 1946.

Comis, C. D., & Leng., G. Action of putative neurotransmitters in guinea pig cochlea. *Experimental Brain Research*, 1979, *36*, 119–128.

Dallos, P. *The auditory periphery*. New York: Academic Press, 1973.

Dallos, P., Billone, M. C., Durrant, J. D., Wang, C. Y., & Raynor, S. Cochlear inner and outer haircells: Functional differences. *Science*, 1972, *177*, 356–358.

Dallos, P., & Cheatham, M. A. Production of cochlear potentials by inner and outer hair cells. *Journal of the Acoustical Society of America*, 1976, *60*, 510–512.

Dallos, P., Harris, D., Özdamar, O., & Ryan, A. Behavioral, compound action potentials and single unit threshold: Relationship in normal and abnormal ears. *Journal of the Acoustical Society of America*, 1978, *64*, 151–157.

Davis, H. Biophysics and physiology of the inner ear. *Physiological Reviews*, 1957, *37*, 1–49.

Davis, H. A model for transducer action in the cochlea. *Cold Spring Harbor Symposia on Quantitative Biology*, 1965, *30*, 181–190.

Davis, H., Benson, R. W., Covell, W. P., Fernandez, C., Goldstein, R., Katsuki, Y., Legouix, J. P., McAuliffe, D. R., & Tasaki, I. Acoustic trauma in the guinea pig. *Journal of the Acoustical Society of America*, 1953, *25*, 1180–1189.

Davis, H., Fernandez, C., & McAuliffe, D. R. The excitatory process in the cochlea. *Proceedings of the National Academy of Sciences* (U.S.A.), 1950, *36*, 580–587.

Djupesland, G., & Zwislocki, J. J. The effect of temporal summation on the human stapedius reflex. *Acta Otolaryngologica* (Stockholm) 1971, *71*, 262–265.

Djupesland G., & Zwislocki, J. J. On the critical band in acoustic reflex. *Journal of the Acoustical Society of America*, 1973, *54*, 1157–1159.

Duifhuis, H. Cochlear linearity and second filter: Possible mechanism and applications. *Journal of the Acoustical Society of America*, 1976, *59*, 408–423.

Eldredge, D. H. A role of the cochlear microphonic: Suppression of auditory action potentials. In S. K. Hirsh, D. H. Eldrege, I. J. Hirsh, & S. R. Silverman (Eds.), *Hearing and Davis*. St. Louis, Missouri: Washington University Press, 1976.

Engström, H., & Engström, B. Structure of the hairs on cochlear sensory cells. *Hearing Research*, 1978, *1*, 49–66.

Feddersen, W. E., Sandel, T. T., Teas, D. C., & Jeffress, L. A. Localization of high-frequency tones. *Journal of the Acoustical Society of America*, 1957, *29*, 988–991.

Fex, J. Auditory activity in centrifugal and centripetal fibers in the cat. A study of a feedback system. *Acta Physiologica Scandinavica* (Stockholm) 1962, *55*, Suppl. 189.

Fex, J. Neuropharmacology and potentials in the inner ear. In A. R. Møller (Ed.), *Basic mechanisms in hearing*. New York: Academic Press, 1973.

Firestone, F. A. The phase difference and amplitude ratio at the ears due to a source of pure tone. *Journal of the Acoustical Society of America*, 1930, *2*, 260–270.

Flock, A. Electron microscopic and electrophysiological studies on the lateral line organ. *Acta Otolaryngologica* (Stockholm), 1965 (Suppl. *199*), 1–90. a

Flock, A. Transducing mechanisms in lateral line canal organ receptors. *Cold Spring Harbor Symposia on Quantitative Biology*, 1965, *30*, 133–146. b

Flock, A. Sensory transduction in hair cells. In W. Loewenstein (Ed.), *Handbook of sensory physiology* (Vol. 1). Pp. 396–441. Berlin: Springer-Verlag, 1971.

Flock, A. Contractible proteins in hair cells. *Hearing Research*, 1980, *2*, 411–412.

Flock A., & Cheng, H. C. Actin filaments in sensory hairs of inner ear receptor cells. *Journal of Cell Biology*, 1977, *75*, 339–343.

Flock, A., Flock, B., & Murray, E. Studies on the sensory hairs of receptor cells in the inner ear. *Acta Otolaryngologica* (Stockholm), 1977, *83*, 85–91.

Flock, A., & Lam, D.M.K. Neurotransmitter synthesis in inner ear and lateral sense organs. *Nature* (London), 1974, *249*, 142–144.

Fumagalli, Z. Ricerche morfologiche sull'apparato di transmissione del suono (Sound conducting apparatus: A study of morphology). *Archivio Italiano di Otologia, Rinologia e Laringologia*, 1949, *60*,Suppl. 1.

Godfrey, D. A., Carter, J. A., Berger, S. J., & Matchinsky, F. M. Levels of putative transmitter amino acids in the guinea pig cochlea. *Journal of Histochemistry and Cytochemistry*, 1976, *24*, 468–470.

Gray, A. A. On a modification of the Helmholtz theory of hearing. *Journal of Anatomy and Physiology* (London) 1900, *34*, 324–350.

Guinan, J. J., & Peake, W. T. Middle-ear characteristics of anesthetized cats. *Journal of the Acoustical Society of America*, 1967, *41*, 1237–1261.

Hartley, R. V. L., & Frey, T. C. The binaural localization of pure tones. *Physiological Reviews*, 1921, *18*, 431–442.

Helmholtz, H. L. F. (1) Die Lehre von den Tonempfindungen als physiologische Grundlage für die Theorie der Musik. Brunswick, Germany: Vieweg-Verlag, 1863.

Henson, O. W., Jr. The activity and function of the middle ear muscles in echolocating bats. *Journal of Physiology*, (London) 1965, *180*, 871–887.

Johnstone, B. M., & Boyle, A. J. F. Basilar membrane vibration, examined with the Mössbauer technique. *Science*, 1967, *158*, 389–390.

Johnstone, B. M., & Taylor, K. M. Mechanical aspects of cochlear function. In R. Plomp & G. F. Smoorenburg (Eds.), *Frequency analysis and periodicity detection in hearing*. Leiden: A. W. Sijthoff, 1970.

Johnstone, B. M., Taylor, K. M., & Boyle, A. J. Mechanics of the guinea pig cochlea. *Journal of the Acoustical Society of America*, 1970, *47*, 504–509.

Kiang, N.-Y. S., & Peake, W. T. Components of electrical responses recorded from the cochlea. *Annals of Otology, Rhinology, and Laryngology*, 1960, *69*, 448–458.

Kimura, R. Hairs of the cochlear sensory cells and their attachment to the tectorial membrane. *Acta Otolaryngologica* (Stockholm), 1965, *61*, 55–72.

Kinsler, L. E., & Frey, A. R. *Fundamentals of acoustics* (2nd ed.). New York-London: Wiley, 1962.

Klinke R., & Galley, N. Efferent innervation of vestibular and auditory receptors. *Physiological Reviews*, 1974, *54*, 316–357.

Klinke, R., & Oertell, W. Amino acids—putative afferent transmitter in the cochlea? *Experimental Brain Research*, 1977, *30*, 145–148.

Konishi, T., & Nielsen, D. W. The temporal relationship between basilar membrane motion and nerve impulse initiation in auditory nerve fibers of guinea pigs. *Japanese Journal of Physiology,* 1978, *28,* 291-307.

Lim, D. Morphological relationship between the tectorial membrane and the organ of Corti. Scanning and transmission electron microscopy. *Journal of the Acoustical Society of America,* 1971, *50,* 92(A).

Lim, D. J. Fine morphology of the tectorial membrane. *Archives of Otolaryngology,* 1972, *96,* 199-215.

Melloni, B. J. The internal ear.*What's new?* 1957, *57,* 14-19.

Møller, A. R. Intra-aural muscle contraction in man, examined by measuring acoustic impedance of the ear. *Laryngoscope,* 1958, *68,* 48-62.

Møller, A. R. Network model of the middle ear. *Journal of the Acoustical Society of America,* 1961, *33,* 168-176. a

Møller, A. R. Bilateral contraction of the tympanic muscles in man. *Annals of Otology, Rhinology, and Laryngology,* 1961, *70,* 735-752. b

Møller, A. R. Acoustic reflex in man. *Journal of the Acoustical Society of America,* 1962, *34,* 1524-1534. a

Møller, A. R. The sensitivity of contraction of the tympanic muscles in man. *Annals of Otology, Rhinology, and Laryngology,* 1962, *71,* 89-95. b

Møller, A. R. Transfer function of the middle ear. *Journal of the Acoustical Society of America,* 1963, *35,* 1526-1534.

Møller, A. R. The acoustic impedance in experimental studies on the middle ear. *International Audiology,* 1964, *3,* 123-135. a

Møller, A. R. Effect of tympanic muscle activity on movement of the eardrum, acoustic impedance and cochlear microphonics. *Acta Otolaryngologica* (Stockholm), 1964, *58,* 525-534. b

Møller, A. R. An experimental study of the acoustic impedance of the middle ear and its transmission properties. *Acta Otolaryngologica* (Stockholm), 1965, *60,* 129-149.

Møller, A. R. The middle ear. In J. V. Tobias (Ed.), *Foundations of modern auditory theory.* New York: Academic Press, 1972.

Møller, A. R. The acoustic middle ear muscle reflex. In W. D. Keidel & W. D. Neff (Eds.), *Handbook of sensory physiology* (Vol. V/I). New York: Springer-Verlag, 1974.

Olson, H. F. *Dynamical analogies.* Princeton: D. Van Nostrand Co., 1958.

Peake, W. T., & Kiang, N.-Y. S. Cochlear responses to condensation and rarefaction clicks. *Biophysical Journal,* 1962, *2,* 23-34.

Pierson, M., & Møller, A. R. Some dualistic properties of the cochlear microphonics. *Hearing Research,* 1980, *2,* 135-149. a

Pierson, M., & Møller, A. R. Effect of modulation of basilar membrane position on the cochlear microphonic. *Hearing Research,* 1980, *2,* 151-162. b

Popelka, G. R., Margolis, R. H., & Wiley, T. L. Effect of activating signal bandwidth on acoustic reflex thresholds. *Journal of the Acoustical Society of America,* 1976, *59,* 153-159.

Rasmussen, H. Studies on the effect on the air conduction and bone conduction from changes in meatal pressure in normal subjects and otosclerotic patients. *Acta Otolaryngologica* (Stockholm), 1948 (Suppl.), *74,* 54-64.

Rhode, W. S. Observations of the vibration of the basilar membrane in squirrel monkeys using Mössbauer technique. *Journal of the Acoustical Society of America,* 1971, *49,* 1218-1231.

Rhode, W. S. An investigation of post-mortem cochlear mechanics using the Mössbauer effect. In A. R. Møller (Ed.), *Basic mechanisms in hearing.* New York: Academic Press, 1973.

Rhode, W. S. Some observations on cochlear mechanics. *Journal of the Acoustical Society of America,* 1978, *64,* 158–176.

Rhode, W. S., & Robles, L. Evidence from Mössbauer experiments for nonlinear vibration in the cochlea. *Journal of the Acoustical Society of America,* 1974, *55,* 588–596.

Robles, L., Rhode, W. S., & Geisler, C. D. Transient response of the basilar membrane measured in squirrel monkeys using Mössbauer effect. *Journal of the Acoustical Society of America,* 1976, *59,* 926–939.

Russell, I. J., & Sellick, P. M. Intracellular studies of hair cells in the mammalian cochlea. *Journal of Physiology* (London), 1978, *284,* 261–290.

Russell, I. J., & Sellick, P. M. The responses of hair cells to low frequency tones and their relationship to extracellular receptor potentials and sound pressure level in the guinea pig cochlea. In J. Syka & L. Aitkin (Eds.), *Proceedings of the Satellite Symposium to the XXVIII International Congress of Physiological Sciences.* New York: Plenum Press, 1981, 3–15.

Sellick, P. M., & Russell, I. J. The responses of inner hair cells to basilar membrane velocity during low frequency auditory stimulation in the guinea pig cochlea. *Hearing Research,* 1980, *2,* 439–445.

Shaw, E. A. C. Transformation of sound pressure level from the free field to the eardrum in the horizontal plane. *Journal of the Acoustical Society of America,* 1974, *56,* 1848–1861. a

Shaw, E. A. C. The external ear. In W. D. Keidel & W. D. Neff (Eds.), *Handbook of sensory physiology* (Vol. V/1). 1974, New York: Springer-Verlag. b

Simmons, F. B. Middle ear muscle activity at moderate sound levels. *Annals of Otology, Rhinology and Laryngology,* 1959, *68,* 1126–1143.

Smith, C. A. Preliminary observations on the terminal ramifications of nerve fibers in the cochlea. *Acta Otolaryngologica* (Stockholm), 1973.

Sokolich, W., Hamernik, R. P., Zwislocki, J. J., & Schmiedt, R. A. Inferred response polarities of cochlear hair cells. *Journal of the Acoustical Society of America,* 1976, *59,* 963–974.

Spoendlin, H. Innervation pattern of the organ of Corti of the cat. *Acta Otolaryngologica* (Stockholm), 1969, *67,* 239–254.

Spoendlin, H. Structural basis of peripheral frequency analysis. In R. Plomp & G. F. Smoorenburg (Eds.), *Frequency analysis and periodicity detection in hearing.* Leiden: A. W. Sijthoff, 1970.

Spoendlin, H. The innervation of the cochlear receptor. In A. R. Møller (Ed.), *Basic mechanism of hearing.* New York: Academic Press, 1973.

Stuhlman, O. The nonlinear transmission characteristics of the auditory ossicles. *Journal of the Acoustical Society of America,* 1937, *9,* 119–128.

Stuhlman, O. *An introduction to biophysics.* New York: Wiley, 1943.

Teranishi, R. & Shaw, E.A.G. External ear acoustic models with simple geometry. *Journal of the Acoustical Society of America,* 1968, *44,* 257–263.

Thornhill, R. A. Biochemical and histochemical studies on vestibular neurotransmission. In Symposium Mechanoreception. *Abhandlungen der Rheinish-Westfälischen Akademic der Wissenschaften,* 1974, *53,* 209–221.

Tonndorf, J., & Khanna, S. M. The role of the tympanic membrane in middle ear transmission. *Annals of Otology, Rhinology and Laryngology,* 1970, *79,* 743–754.

Tonndorf, J., Khanna, S. M., & Fingerhood, B. The input impedance of the inner ear in cats. *Annals of Otology, Rhinology and Laryngology,* 1966, *75,* 752–763.

Tonndorf, J., McCardle, F., & Kruger, B. Middle ear transmission losses caused by tympanic membrane perforations in cats. *Acta Otolaryngologica* (Stockholm) 1976, *81,* 330–336.

Wever, E. G., & Bray, C. W. The nature of acoustic response: The relation between sound intensity and magnitude of responses in the cochlea. *Journal of Experimental Psychology,* 1936, *19,* 129–143.

Wever, E. G., & Lawrence, M. *Physicological acoustics.* Princeton, New Jersey: Princeton University Press, 1954.

Wever, E. G., Lawrence, M., & Smith, K. R. The middle ear in sound conduction. *Archives of Otolaryngology,* 1948, *48,* 19–35.

Whitfield, I. C., & Ross, H. F. Cochlear-microphonic and summating potentials and the output of individual hair-cells generators. *Journal of the Acoustical Society of America,* 1965, *38,* 126–131.

Wiederhold, M. L. Variations in the effects of electrical stimulation of the crossed olivocochlear bundle on cat single auditory nerve fiber response to tone bursts. *Journal of the Acoustical Society of America,* 1970, *48,* 966–977.

Wiederhold, M. L., & Kiang, N.-Y. S. Effects of electrical stimulation of the crossed olivocochlear bundle on single auditory nerve fibers in the cat. *Journal of the Acoustical Society of America,* 1970, *48,* 950–965.

Wilson, J. P., & Johnstone, J. R. Capacitive probe measures of basilar membrane vibration. In *Hearing theory.* The Netherlands: IPO Eindhoven, 1972.

Wilson, J. P., & Johnstone, J. R. Basilar membrane and middle-ear vibration in guinea pig measured by capacitive probe. *Journal of the Acoustical Society of America,* 1975, *57,* 705–723.

Zakrisson, J. E. The role of the stapedius reflex in post-stimulatory auditory fatigue. *Acta Otoloryngologica* (Stockholm), 1975, *79,* 1–10.

Zweig, G., Lipes, R., & Pierce, J. R. The cochlear compromise. *Journal of the Acoustical Society of America,* 1976, *59,* 975–982.

Zwislocki, J. J. Some impedance measurements on normal and pathological ears. *Journal of the Acoustical Society of America,* 1957, *29,* 1312–1317.

Zwislocki, J. J. Analysis of the middle ear function. Part II. Guina-pig ear. *Journal of the Acoustical Society of America,,* 1963, *35,* 1034–1040.

Zwislocki, J. J. A possible neuro-mechanical sound analysis in the cochlea. *Acustica,* 1974, *31,* 354–359.

Zwislocki, J. J. Five decades of research on cochlear mechanics. *Journal of the Acoustical Society of America,* 1980, *67,* 1679–1685.

Zwislocki, J. J., & Sokolich, W. G. Evidence for phase opposition between inner and outer haircells. *Journal of the Acoustical Society of America,* 1974, *55,* 466(A).

Zwislocki-Moscicki, J. J. Theorie der Schneckenmechanik. Qualitative und quantitative Analyse. *Acta Otolaryngologica* (Stockholm), Suppl. 72 1948, 1–76.

CHAPTER 2

Anatomy and General Function
of the Auditory Nervous System

Introduction

The ascending pathway of the auditory system is more complex than that of other sensory systems. For example, studies of the neural mechanism in somesthesia (tactile stimuli), sensed by various receptors in the skin, are related to the cerebral cortex via a three-neuron pathway (see Mountcastle, 1974). The organization of each of these stations is simple and the information is preserved much in its original form when it reaches the sensory cortex. In the auditory system, however, some channels pass via four nuclei, whereas others pass through five nuclei before reaching the auditory part of the cerebral cortex. Additionally, various nuclei in the ascending auditory pathway are more complex than those in most other ascending sensory systems, and there are several connections between the pathways of the left and right sides.

Most of our knowledge about the function of the nervous system is based on results obtained by using neurophysiological methods together with morphological methods. Such studies are performed in experimental animals, and the response patterns to different types of sounds are then studied. Neurophysiological methods used to study information processing in the auditory system are based on recordings of the discharge pattern in single nerve cells and nerve fibers. Recordings performed so far have, for the most part, been extracellular recordings using various types of microelectrodes.

Transformation of information that takes place in the ascending auditory pathway occurs in the various brain nuclei of this pathway. In these nuclei, afferent nerve fibers from lower centers are received at synaptic terminals on nerve cells that, in turn, send axons to higher centers. Since input to these neurons is both inhibitory and excitatory, certain channels can inhibit or attenuate the transmission of information in other channels. Since many nerve fibers terminate on each nerve cell, information from many channels is integrated. The complexity of this network is illustrated by Figure 2.1, where a single nerve cell in the cochlear nucleus is represented. In addition, there are abundant interconnections between nerve cells and feedback connections whereby, certain parts of the sensory-neural network may perform various filtering functions. Interplay between excitation and inhibition in such networks is assumed to be an important part of their function.

As far as transformation of information in the ascending sensory pathways is concerned, recent work has focused on the *coding* of information in the discharge pattern of single fibers or nerve cells. In this context, coding usually means the way certain characteristics of a particular stimulus are represented in the discharge pattern of a single nerve cell or fiber. One aim of this work is to determine characteristics of the stimuli to which the various neurons respond and to elucidate the neural organization responsible for this preference. In other words, the aim is to obtain a description of the "circuit diagram" of the neural network that accomplishes a certain transformation of the stimulus. Such work may be regarded as being relatively independent of studies of the function of individual neurons, which may be seen as "building blocks." These building blocks may have similar properties, although they may perform completely different tasks depending on the circuit of which they are a part.

When it first was possible to study systematically the responses from single auditory nerve fibers, it became apparent that the most prominent feature of these responses was frequency selectivity. Individual nerve fibers responded to pure tones in such a way that a particular fiber had its lowest threshold at a certain frequency, and changing the frequency up or down always increased the threshold. Similar, although quantitatively different, frequency tuning could be seen in neurons at the various brain nuclei that constituted the ascending auditory pathway.

These data have been interpreted very differently, and frequency tuning may not play the same fundamental role when *natural* sounds are concerned. Frequency selectivity of the ear will be discussed in the following chapters and treated in more detail in Chapter 3.

The complexity of methods used in analysis of the recorded discharge pattern has increased since the first recordings using microelectrodes were obtained. At that time, discharges were displayed on an oscilloscope and

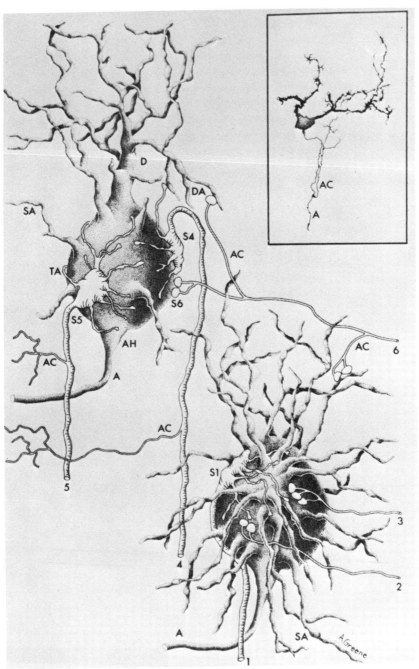

FIGURE 2.1. Schematic drawing of a single nerve cell in the interstitial and anteroventral cochlear nucleus of a cat. (From Morest *et al.*, 1973.)

photographed. Threshold of the stimulus evoking a change in the discharge rate was often the parameter that was studied. When computers became generally available and their capacity increased, statistical signal analysis techniques were applied to a greater and greater extent to the analysis of the discharge pattern. At the same time, more complex stimuli were used in place of the pure tones and click sounds that constituted almost the only stimuli used in earlier phases of auditory neurophysiology. At present, some type of statistical signal analysis is invariably applied to the recorded impulse trains to describe their characteristics (distribution of intervals, etc.) and their relationship to the stimulus (see Marmarelis & Marmarelis, 1978).

Information from different fibers of the auditory nerve is combined and controlled by the neural integration that occurs in the various brain nuclei. These transformations are presently the object of studies using microelectrode recordings and statistical signal analysis methods, and results from different levels of the ascending sensory pathway have been compared.

Much work has been aimed at classifying cells in the various brain nuclei according to their response patterns. Each of the various nuclei may be considered to function as a filter, suppressing certain characteristics of the information and enhancing others. As an example, adaptation may be regarded as suppressing slowly varying sounds and enhancing rapidly varying sounds. This phenomenon allows for optimal use of the available afferent signal, since steady or slowly varying sounds really do not carry much information, whereas rapidly varying sounds are the main information-bearing elements in most naturally occurring sound, as, for example, in speech sounds.

In the following, the coding of information in the ascending auditory pathway, and in its different relay stations, will be considered from anatomic and neurophysiological viewpoints.

Anatomical Organization of the Ascending Auditory Pathway

A simplified scheme of the ascending auditory pathway as it is presently understood is shown in Figure 2.2. Fibers of the auditory nerve terminate on cells in both the ventral and dorsal parts of the cochlear nucleus. They in turn, send axons along three fiber tracts to higher centers, the corpus trapezoideum, the (intermediate) stria of Held, and the (dorsal) stria of Monakow. All three tracts project onto the contralateral inferior colliculus through the lateral lemniscus. The trapezoidal tract and

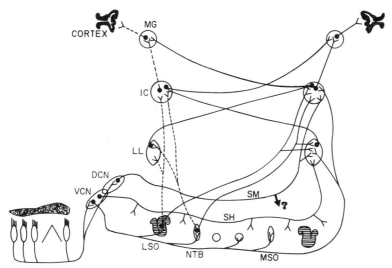

FIGURE 2.2. Ascending auditory pathways. DCN, dorsal cochlear nucleus; VCN, ventral cochlear nucleus; LSO, S-segment of superior olive; MSO, medial superior olive nucleus; NTB, nucleus of the trapezoidal body; LL, lateral lemniscus; IC, inferior colliculus; MG, medial geniculate; SH, stria of Held; SM, stria of Monakow.

the stria of Held connect to neurons in the superior olive complex and the nucleus of the trapezoidal body. The lateral and medial superior olives also connect to the ipsilateral lateral lemniscus and to the inferior colliculus. The lateral lemniscus carries the information to the midbrain relay station, the inferior colliculus. From here, fibers run to the nucleus of the medial geniculate which then sends *its* fibers to terminate in the primary receiving area of the auditory cortex. There are connections between the ascending pathways of the two sides on at least two levels, in the superior olive and in the inferior colliculus. Both are assumed to be of importance in directional hearing. In addition, more diffuse pathways, probably running through more medial brainstem structures, lead from the cochlea to the cortex. The acoustic middle ear reflex receives its input from the ascending auditory pathway at the level of the superior olive nucleus (see p. 39).

AUDITORY NERVE

The *auditory nerve* is the auditory part of the eighth cranial nerve, which also embodies the vestibular nerve. The auditory nerve consists of the axons originating from the afferent terminals of both inner and outer hair cells. As mentioned previously (see p. 72), approximately 95% of the fibers in the auditory nerve originate from inner hair cells and only 5% from outer hair cells. The former connect to one or a few hair cells while the latter (Figure 1.57) probably represent a longer segment of the basilar membrane (Spoendlin, 1970).

The fibers of the auditory nerve maintain the orderly organization of the sensory epithelium so that fibers originating in the apex of the cochlea are found in the center of the auditory nerve, and nerve fibers that originate at the base of the cochlea are found in the periphery of the auditory nerve (Sando, 1965).

There are approximately 30,000 nerve fibers in the human auditory nerve. The cat and the guinea pig, two animals often used in auditory research, possess about 50,000 and 25,000 nerve fibers, respectively. The monkey has about 31,000 nerve fibers (Gacek & Rasmussen, 1961). The porpoise and the whale have over 100,000 nerve fibers in each auditory nerve. Fiber diameters range between 1 and 8μ, with most fibers having diameters of from 3 to 6μ. In man, the auditory nerve has approximately 30,000 fibers (Larsell, 1951). The fiber diameters in man are about the same as these in guinea pigs and cats, with the majority being from 2 to 4μ in diameter (Lazorthes, Lacomme, Gaubert & Planel, 1961).

Cell bodies of these nerve cells constitute the spiral ganglion located in the spiral osseus. Each nerve fiber is unmyelinated at first, acquiring a myelin sheath only after passing through the holes of the habenula perforata. It is likely that nerve impulses are set up in the first node of Ranvier at the beginning of the myelinated portion. It has been shown that there are large cells scattered along the auditory nerve (Harrison & Warr, 1962). These cells receive synaptic input from collaterals of auditory nerve fibers. (These cells are not the same as the interstitial nucleus described by Lórente de No, 1933b.)

COCHLEAR NUCLEUS

The *cochlear nucleus* is a complex nucleus, possessing many different types of neurons, some of which represent the first relay stations for auditory information. Morphologically, the cochlear nucleus is often divided into three major divisions: the anterior ventral (AVCN), the posterior ventral (PVCN), and the dorsal (DCN) cochlear nucleus. The auditory nerve enters the brainstem and bifurcates into an ascending branch that projects to the ventral cochlear nucleus (VCN), and a descending branch that innervates the PVCN and the DCN. There are also other sources of innervation of the DCN. One is probably a continuation of the descending branch of the auditory nerve and another consists of fibers from the AVCN. The DCN also receives connections from the contralateral ascending auditory pathway. Cells in the AVCN and PVCN are rather small and receive relatively few (3–10) nerve fibers each, whereas the cells in the DCN are large and receive input from as many as 100 or more primary fibers (Lórente de No, 1933a, b; Osen, 1969; Kane, 1978; Harrison & Irving, 1966; Harrison & Warr, 1962). Figure 2.3 shows a schematic outline of the cochlear nucleus complex.

There are three main pathways *from* the cells in the cochlear nucleus to higher centers—namely, the dorsal ascending pathway (stria of Monakow), the intermidiate ascending pathway (stria of Held), and the ventral stria (trapezoid body) (van Noort, 1969; Harrison & Feldman, 1970). Fibers from the neurons in the trapezoidal body connect ipsilaterally to the lateral superior olive (LSO) nucleus, bilaterally to the medial superior olive (MSO), and contralaterally to the nucleus of the trapezoidal body. They then ascend in the contralateral lateral lemniscus (LL) to the inferior colliculus (IC). The stria of Held originates in cells in the posterior ventral cochlear nucleus (Harrison & Irving, 1966; Warr, 1969; Osen, 1969; van Noort, 1969) and terminates in the periolivary nuclei of both sides. The stria of Monakow has been studied less extensively, but seems to originate in the dorsal cochlear nucleus (DCN) and possibly connects to the con-

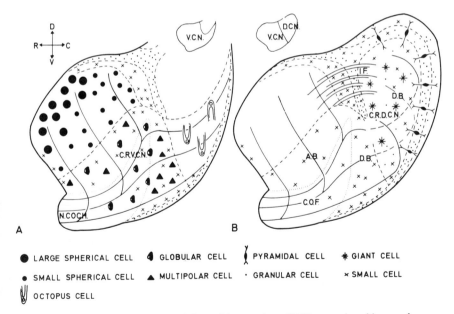

● LARGE SPHERICAL CELL   ◀ GLOBULAR CELL   ❘ PYRAMIDAL CELL   ✳ GIANT CELL

• SMALL SPHERICAL CELL   ▲ MULTIPOLAR CELL   · GRANULAR CELL   × SMALL CELL

Ⓤ OCTOPUS CELL

FIGURE 2.3. Schematic outline of the cochlear nucleus. VCN, ventral cochlear nucleus; DCN, dorsal cochlear nucleus. (From Osen, 1969.)

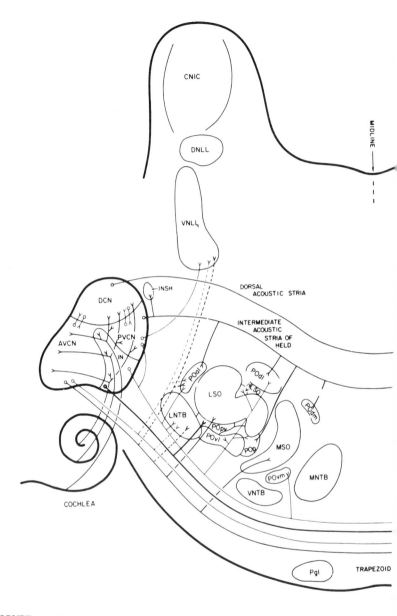

FIGURE 2.4. Schematic drawing of the known projections of the cochlear nucleus. The cochlea is represented by a spiral, the cochlear nucleus by an idealized sagittal section, the medulla by an idealized transverse section, and the inferior colliculus by an idealized transverse section in a more rostral plane. Solid lines, pathways considered to be well documented; dotted lines, pathways for which some evidence exists, but not of a conclusive nature. To simplify the drawing, the pathways are not strictly correct in all anatomic details (such as the relative position of various fiber components in the trapezoid body). AVCN, anteroventral cochlear nucleus; CNIC, central nucleus of the inferior colliculus; DCN, dorsal cochlear nucleus; DNLL, dorsal nucleus of the lateral lemniscus; HSLO, dorsal hilus of

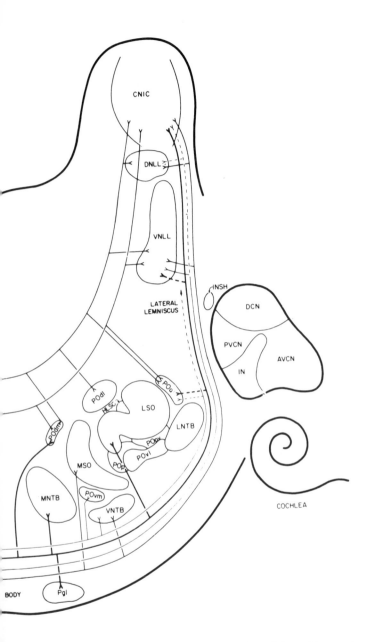

the lateral superior olivary nucleus; IN, interstitial nucleus of the cochlear nucleus; INSH, interstitial nucleus of the stria of Held; LNTB, lateral nucleus of the trapezoid body; LSO, lateral superior olivary nucleus; MNTB, medial nucleus of the trapezoid body; MSO, medial superior olivary nucleus; Pgl, lateral paragigantocellular nucleus; POal, anterolateral periolivary nucleus; POdl, dorsolateral periolivary nucleus; POdm, dorsomedial periolivary nucleus; POp, posterior periolivary nucleus; POpv, posteroventral periolivary nucleus; POvl, ventrolateral periolivary nucleus; POvm, ventromedial periolivary nucleus; PVCN, posteroventral cochlear nucleus; VNLL, ventral nucleus of the lateral lemniscus; VNTB, ventral nucleus of the trapezoid body. (From Kiang, 1975.)

tralateral superior olive (Fernandez & Karapas, 1967) and after which it reaches the contralateral inferior colliculus (van Noort, 1969). The dorsal-most surface layer of cells has been called the fusiform layer (Osen, 1969, 1970; Koningsmark, 1973; Kiang, Morest, Godfrey, Guinan, & Kane, 1973). A schematic outline of the cochlear nucleus is shown in Figure 2.3, and a detailed scheme of the connections from the cochlear nucleus is shown in Figure 2.4.

SUPERIOR OLIVE COMPLEX

The *superior olivary complex* (SOC) consists of the medial superior olive (MSO), lateral superior olive (LSO), a segment, and nucleus of the trapezoidal body (NTB). The SOC exhibits considerable morphologic differences among species. Cells in this complex are the lowest in the ascending auditory pathway to receive input from both ears. The LSO receives its innervation from the ipsilateral AVCN (Harrison & Warr, 1962; Warr, 1966). The MSO receives an almost symmetrical projection from the ipsilateral and contralateral cochlear nuclei (Figure 2.4), and is assumed to play an important role in directional hearing (see p. 174). All fibers that reach the MSO pass through the trapezoidal body and may be collaterals of fibers that travel from the trapezoidal body to the LSO (Warr, 1966). The medial nucleus of the trapezoidal body (MNTB) is also regarded as part of the superior olivary complex. It receives its input predominantly from the contralateral cochlear nucleus (Jungert, 1958; Stotler, 1953; Koningsmark, 1973; Morest, Kiang, Kane, Guinan, & Godfrey, 1973). The lateral nucleus of the trapezoid body (LNTB) receives its input from the ipsilateral interstitial nucleus (Harrison & Feldman, 1970). There is also contralateral input from the PVCN. The VNTB receives its afferent input from the contralateral AVCN (Harrison & Feldman, 1970), but may also receive ipsilateral input.

The fibers from the cells in the superior olive are connected for the most part in the ipsilateral lateral lemniscus, which also receives direct connections from the ipsilateral cochlear nucleus, which bypasses the superior olive neurons.

LATERAL LEMNISCUS

The *lateral lemniscus* (LL) consists of two distinct sets of cells: the dorsal (LLD) and ventral (LLV) groups. The LLV receives its major input from the contralateral cochlear nucleus along with a few fibers from the ipsilateral cochlear nucleus (van Noort, 1969). The LLD receives input exclusively from the contralateral cochlear nucleus, whereas binaural input may come to this nucleus from both the contralateral LSO and the ipsilateral MSO.

INFERIOR COLLICULUS

The *inferior colliculus* is composed of a central nucleus (CNIC), the external nucleus (ENIC), and the pericentral nucleus (PCNIC; Rockel & Jones, 1973a, b, c; Morest, 1964a, 1966a, b; van Noort, 1969). Information reaches the inferior colliculus from the contralateral ventral cochlear nucleus via the trapezoidal tract, and from the dorsal cochlear nucleus via the stria of Monakow. The inferior colliculus also receives input from the ipsilateral ventral cochlear nucleus via a relay in the superior olive nucleus. This route passes through the contralateral superior colliculus as well. Some of the connections to the inferior colliculus are via the lateral lemniscus. Most fibers leaving the inferior colliculus terminate in the ipsilateral medial geniculate nucleus (see van Noort, 1969).

MEDIAL GENICULATE

The *medial geniculate* (MG) is the thalamic auditory nucleus. The neurons of the medial geniculate ganglion receive almost all their input from the ipsilateral inferior colliculus through the brachium of the inferior colliculus. In the cat, the MG contains approximately 250,000 cells (Chow, 1951). Two structurally distinct regions of the MG—a principal cell division (MGP) and a large cell division (MGM), also known as pars principalis and pars magnocellularis (Erulkar, 1975; Moore & Goldberg, 1963) have been identified. Two different cells types have been identified (Morest, 1964b; Rose, 1949), and in the MG, Morest (1964b) showed that there were three major cytoarchitectural divisions—the dorsal, ventral, and medial divisions. The ventral division seems to be the primary afferent relay. The neurons of the MG project onto the primary auditory receiving area.

AUDITORY CORTEX

The auditory cortex is the part of the cerebral cortex that is assumed to be directly associated with auditory information. Many studies have concerned mappings of projections from lower auditory centers to the principal auditory field AAF. Maps of the different areas differ greatly from investigator to investigator. Figure 2.5 shows mappings from two different animals (Merzenich, 1980).

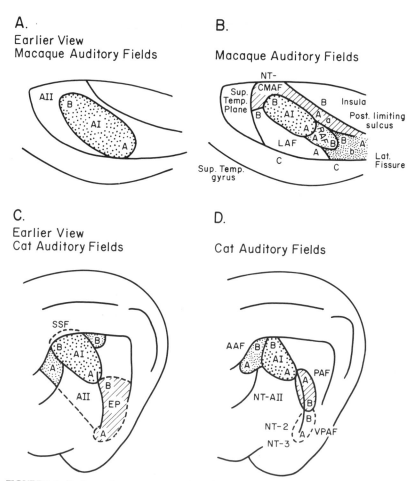

FIGURE 2.5. Earlier and contemporary view of extension of the primary auditory cortex in different animal species. (From Merzenich, 1980.)

The *primary auditory cortex* (AI) receives its input from the anterior and medial geniculate bodies, whereas the other auditory cortical areas (AII and the insular and temporal areas) receive input predominantly from the posterior part of the medial geniculate body (Woolsey, 1960). In contrast to other results, it seems that not only AI receives direct information from the medial geniculate but adjacent cortical areas receive equally direct input from the thalamic auditory nucleus. It seems, then, that there are several parallel processing channels from the medial geniculate nucleus to the cortex (Merzenich, Roth, Knight, & Colwell, 1977).

Frequency Tuning in Response to Simple Stimuli

This section is devoted mainly to a description of the available data on frequency tuning, as obtained from recordings from single auditory nerve fibers and to some extent from cells in the various nuclei of the ascending auditory pathway.

FREQUENCY TUNING IN THE PERIPHERY IN RESPONSE TO PURE TONES

Frequency tuning has been studied extensively in the periphery of the auditory system where it is most prominent. Figure 2.6 illustrates the tun-

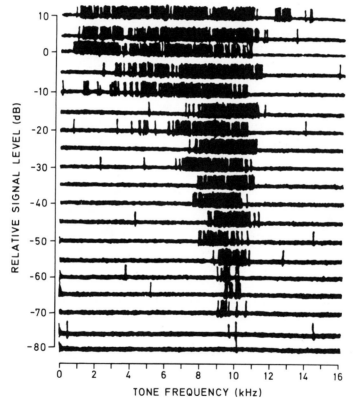

FIGURE 2.6. Frequency sweep method of determination of frequency response area and frequency threshold curve (FTC) of a guinea pig cochlear nerve fiber. A continuous tone is swept linearly in frequency, with 5-dB increments of signal level. Alternate sweeps are in the opposite direction. The pattern of spike discharges is built up on a storage oscilloscope. (Spikes are monophasically positive and .9 mv in amplitude.) Sweep rate: 14 kHz/sec. (From Evans, 1972.)

ing in a single auditory nerve fiber in a guinea pig. Nerve impulses recorded from a single auditory nerve fiber are shown when tones at different levels are presented and their frequency changed as illustrated by the horizontal axis.

It is clear that the fiber only responds to tones within a certain frequency range, and that the frequency range increases with increasing intensity. The threshold has its lowest value at a certain frequency called the unit's characteristic frequency (CF). Threshold increases when the frequency of the stimulus tone, relative to the CF, is either decreased or increased. Thus, in response to pure tones, each fiber or cell is "tuned" to a certain frequency. The area under a frequency threshold curve is called the unit's response area. This frequency tuning is most commonly described quantitatively by means of frequency threshold curves (FTC).[1] It is generally assumed that the origin of tuning is the mechanical frequency selectivity of the basilar membrane. Therefore, different fibers or cells also have different CF, depending on the part of the basilar membrane from which they originate.

In nerve fibers that fire spontaneously, the threshold of firing is usually defined as the first, just-noticeable, increase in the discharge rate (Kiang, Watanabe, Thomas, & Clark, 1965a). The response area of such units is the range of frequencies and intensities within which a tone gives rise to an increase in the discharge rate of a fiber of a cell. Frequency threshold curves thus delimit the area of stimulus intensity and frequency to which a particular fiber responds when stimulated by pure tones, and express the frequency selectivity of a particular nerve fiber. Apart from frequency threshold curves, the frequency selectivity of a single nerve fiber, in response to pure tones, may be studied by using iso-rate and iso-intensity functions. Iso-rate functions are curves showing the stimulus intensity needed to evoke a certain discharge rate as a function of frequency. Iso-intensity functions show the discharge rate as a function of frequency for some constant stimulus intensity (see Geisler, Rhode, & Kennedy, 1974).

Frequency tuning is prominent throughout the ascending auditory pathway, and, consequently, frequency threshold curves and iso-rate and iso-intensity functions may be obtained from nerve cells in the various brain nuclei of the ascending pathway (see Katsuki, Sumi, Uchiyama, &

---

[1] Earlier such functions were called *tuning curves*. In this presentation the term *frequency threshold curves* will be used for curves showing threshold as a function of frequency and the more general term *tuning* (and *tuning curves*) will be used in a broad sense to describe frequency selectivity in general.

Watanabe, 1958, and Katsuki, Suga, & Karmo, 1962). The shapes of these functions show a greater variability from cell to cell than do those of such functions obtained from the fibers of the auditory nerve.

Frequency Threshold Curves

Typical frequency threshold curves of primary auditory nerve fibers of the cat are shown in Figure 2.7. Curves from fibers with different characteristic frequencies (CF) are similar in shape, although those with a low CF are somewhat wider than those with a high CF (Kiang et al., 1965a; Evans, 1972). It is uncertain whether the selectivity of primary nerve fibers may be explained fully on the basis of the mechanical tuning of the basilar membrane, or whether some additional sharpening of the tuning may take place. The existence of two distinct populations of nerve fibers might be expected because some fibers terminate on outer hair cells and others on inner hair cells. Evidence that two such groups of hair cells exist has been presented in a few studies which document a difference in the response pattern obtained from different hair cells (see, e.g., Libermann & Kiang, 1978). However, it is not obvious how these two populations are related to inner and outer hair cells.

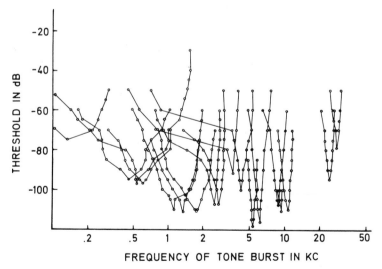

FIGURE 2.7. Typical frequency threshold curves of single auditory nerve fibers in a cat. The curves were obtained using pure tones, presented in 50-msec bursts. (From Kiang et al., 1965a.)

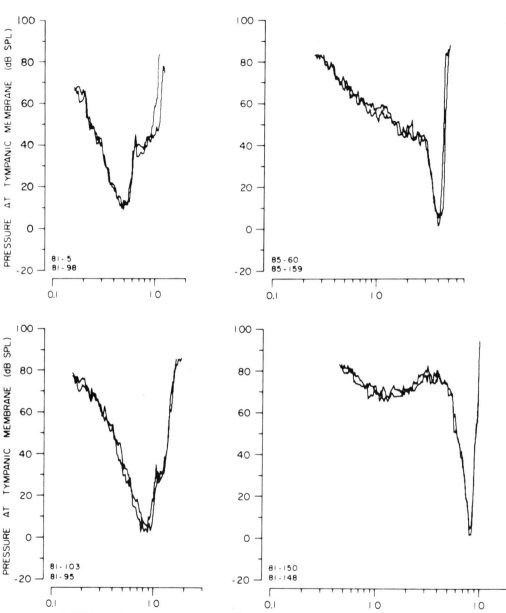

FIGURE 2.8. Frequency tuning curves of six auditory nerve fibers of a cat with different characteristic frequencies. Each panel contains the tuning curves of two different units with similar CF from the same animal. The vertical scale is in decibels SPL at the tympanic membrane. (From Liberman & Kiang, 1978.)

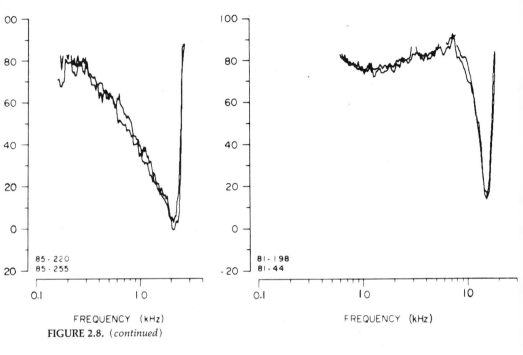

FIGURE 2.8. (*continued*)

Shapes of Tuning Curves

The shape of frequency threshold curves of single auditory nerve fibers varies with the CF in a characteristic way. In fibers with low CF, the frequency threshold curves are nearly symmetrical, whereas for fibers with high CF, they are asymmetrical with a long low frequency "tail" and a steep high frequency portion. That is illustrated in Figure 2.8, where two different regions of frequency threshold curves of primary auditory nerve fibers with high CF can be discerned. One region is the more sensitive region, also called the tip. The other region is the high-threshold portion comprising the frequency range below the tip (Kiang, Sachs, & Peake, 1967). This latter part, also called the tail of the frequency threshold curve, is a nearly flat region that extends toward the low frequencies (Figure 2.8). The threshold in that region is between 40 and 60 dB above the most sensitive region around the fiber's characteristic frequency. The high-frequency part of frequency threshold curves represents a steep increase in threshold to sounds above physiological levels. The presence of the tail portion of the frequency threshold curves implies that a nerve fiber with a high center frequency will respond to low frequency sounds if their intensities are higher than 40–60 dB above the threshold at CF (Kiang

& Moxon, 1974). Most physiologically important sounds have such intensities. Because the tails of the frequency threshold curves are almost flat, response to sounds located in the frequency region of the tail is nearly independent of the frequency; thus, the sound that activates that part of the fiber's response area is not subjected to any spectral filtering in the auditory periphery. Low-frequency sounds (more than 50 dB above threshold) thus evoke responses of two types—one that is spectral specific in fibers tuned to low frequencies, and one that has essentially no spectral specificity and is seen in the fibers that are tuned to high frequencies.

There are reasons to believe that these two regions of the frequency threshold curves represent functionally different types of excitation in the cochlea. It has been shown that the two regions of the frequency threshold curves of primary fibers are influenced differently by anoxia, ototoxic drugs, and masking (Evans & Klinke, 1974; Evans, 1975a, b; Kiang & Moxon, 1974). Figure 2.9 shows how anoxia affects only the tip and leaves the tail of the frequency threshold curve essentially unchanged. Thus, the frequency threshold curves of primary fibers in animals with damaged cochleas attain the shape of lowpass functions with only the tails left unchanged (Evans, 1975a, b). These experimental results have been

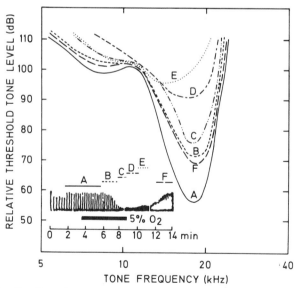

FIGURE 2.9. The effect of anoxia on the frequency threshold curves of an auditory nerve fiber in a guinea pig. (From Evans, 1975a.)

taken to support the hypothesis that in the cochlea there is a physiologically vulnerable second filter in addition to the mechanical filtering that takes place in the cochlea.

However, results of recent work show that the discrepancy between the measured mechanical tuning of the basilar membrane and that of single auditory nerve fibers is likely to be due to the way mechanical tuning has been measured previously. These results show that the selectivity of single auditory nerve fibers can be explained by the mechanical properties of the basilar membrane (Zwislocki & Kletsky, 1979). There is thus no need of a second filter to explain the spectral selectivity of the auditory periphery.

Masking seems to affect the tail portion of the frequency threshold curve very little, while the tip, with proper choice of spectrum and intensity of the masking tone (Kiang & Moxon, 1974), can almost be abolished. Figure 2.10 shows how the tuning curve in a cat is affected by a low-frequency masking noise. The two curves of a single auditory nerve fiber show tuning with and without the masking noise present.

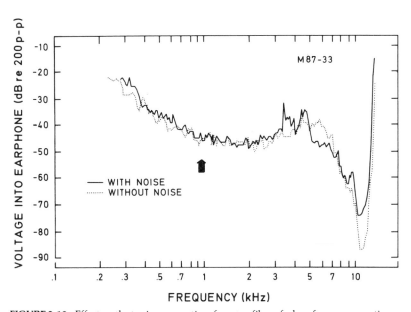

FIGURE 2.10. Effect on the tuning properties of a nerve fiber of a low-frequency continuous noise (arrow) of an intensity just high enough to elicit a response. The bandwidth of the noise was 500 Hz and its level was 45 dB re 70.7 V rms into the earphone. (From Kiang & Moxon, 1974.)

Frequency tuning in the inner ear can also be demonstrated by recording the gross nerve compound action potential (CAP; see p. 94) of the
auditory nerve. This potential may be recorded from several locations in
the inner ear. It is recorded most commonly from the round window. The
CAP originates from the firing of primary fibers, and therefore possesses
frequency-selective properties. The CAP evoked by a weak tone burst or
bandpass-filtered clicks represents the firing of a limited number of fibers
of the auditory nerve. It originates from hair cells in a restricted area of
the basilar membrane—an area where the traveling wave has its maximal
amplitude. Frequency-selective properties of the action potential appear
when the response to such stimuli are masked by a pure tone (Figure 2.11).
The change in threshold of the CAP, evoked by the test stimuli as a func-

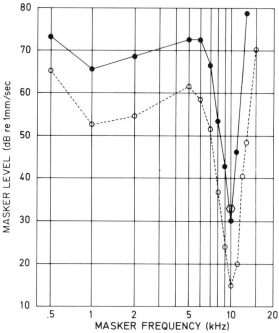

FIGURE 2.11. Masked threshold curves of the compound action potential (CAP) using the
forward method. The CAP was recorded from the round window of a guinea pig and the
sound expressed in velocity of the stapes (dB re 1nm/sec). The stimulus was a 10-kHz tone
with a duration of 15 msec and an intensity of 45 dB. The masker had a duration of 75 msec.
The masker was terminated 7.5 msec before the onset of the stimulus. These open circles and
dashed lines show the level of the masker that decrease the amplitude of the CAP (N_1) by
one-third and the filled circles and solid lines show the masker level that eliminate the CAP.
The hexagonal symbol signifies the frequency and level of the probe tone. (From Dallos &
Cheatham, 1976.)

tion of the frequency of the masking tone, has the shape of a tuning curve
and resembles the frequency threshold curves of single auditory nerve
fibers (Dallos & Cheatham, 1976; Harris, 1979). This method of determining frequency tuning has the practical advantages of being applicable
to studies in humans (Eggermont, 1977), as well as animals, and of
employing experimental techniques that are simpler than recording from
single auditory nerve fibers.

Frequency tuning in response to pure tones is a characteristic of single
neurons in the entire cochlear nucleus complex (Rose, Galambos &
Hughes, 1959). At first glance, the frequency threshold curves obtained in
the cochlear nucleus resemble those obtained from single fibers of the
auditory nerve (Figure 2.12). Further analysis, however, reveals certain
differences (Møller, 1972b). The average width of the frequency threshold
curves of cochlear nucleus units (at 10 dB above threshold at CF) is larger
than that of auditory nerve fibers. This indicates a convergence of
auditory fibers on the nerve cells in the cochlear nucleus, which means
that each cochlear nucleus cell receives more than one auditory nerve
fiber. In fact, many cochlear nucleus cells have hundreds of primary nerve
fibers as input (see Figure 2.1, p. 107). Each cell in the cochlear nucleus

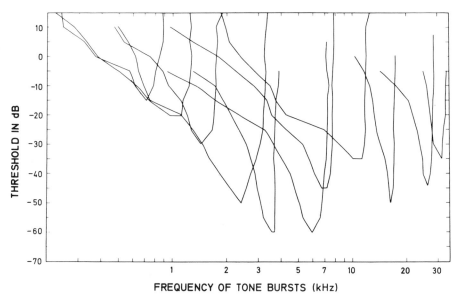

FIGURE 2.12. Typical frequency threshold curves obtained from single nerve cells in the
cochlear nucleus. The vertical scale shows relative level of the sound stimulus in decibels.
(From Møller, 1969a.)

thus represents a greater length of the basilar membrane than does a primary auditory fiber. Whereas the frequency tuning curves of the auditory nerve fibers are rather uniform in shape, except for a slight difference between fibers with high and low CF, tuning curves of cochlear nucleus units vary in shape. Thus there are frequency threshold curves with two peaks (two CF). An example is shown in Figure 2.13.

Many neurons in the superior olive have frequency threshold curves similar to those of the cochlear nucleus neurons and other have frequency threshold curves of a different shape than those of the cochlear nucleus (Guinan, Guinan, & Norris, 1972b; Guinan & Norris, 1972a). Thus, there are neurons with extremely narrow and extremely wide tuning (Guinan *et al.*, 1972b). Examples of different types of tuning curves are shown in Figure 2.14. The average width of frequency tuning curves of neurons in the superior olive is, however, similar to that of cochlear nucleus neurons (Figure 2.12; Tsuchitani & Boudreau, 1967; Boudreau & Tsuchitani, 1970; Tsuchitani, 1977). Figure 2.15, shows the width of frequency threshold curves as a function of center frequency for fibers in the auditory nerve, for neurons in the cochlear nucleus, and for neurons in the superior olive complex. From this graph it follows that the width of the frequency threshold curve increases with center frequency for primary nerve fiber,

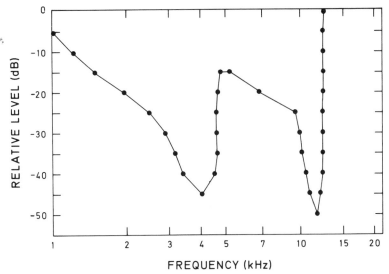

FIGURE 2.13. A frequency tuning curve from a cell in the cochlear nucleus of a rat showing two peaks. The vertical scale shows relative sound level.

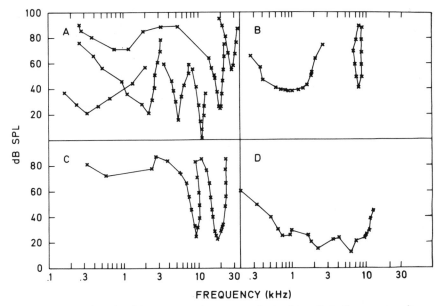

FIGURE 2.14. Sample of frequency tuning curves obtained in cells in the superior olive nucleus. The individual graphs show samples of tuning curves from different classes of units. (Modified from Guinan *et al.*, 1972b.)

cells in the cochlear nucleus, and cells in the superior olive. It is also evident from the figure that the frequency threshold curves in the cochlear nucleus are approximately twice as wide as those in the auditory nerve. It is important to note that the width of the frequency threshold curves of cochlear nucleus neurons is almost identical in the cat and the rat despite anatomical differences in the cochleas of these two animals.

Results shown in Figure 2.15 are only true with regard to the width of the tuning curves. It is also important to consider the other characteristics of the tuning curves. Particular attention has been devoted to the steepness of skirts of the frequency threshold curves. It has been observed that the high frequency skirt of frequency threshold curves of auditory nerve fibers is steeper than the low-frequency skirt, particularly in fibers with CF above 2 kHz. In auditory fibers, the high frequency skirt of the frequency threshold curve may have a slope of 150–600 dB/octave (Evans, 1972; which means that the threshold increases with increasing frequency at a rate of 150–600 dB/octave). At 10 kHz, a slope of 500 dB/octave means that the threshold increases about 1.2 dB for every 10

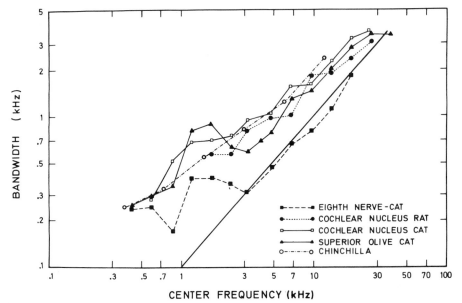

FIGURE 2.15. Widths of frequency threshold curves from different levels of the auditory nervous system and from different animals. The width was measured at 10 dB above the threshold at CF. The curves with symbols show the average values as a function of CF. Each datum point represents the bandwidth values covering half octaves. The straight line without symbols represents the frequency band covered by a .35-mm segment of the basilar membrane in the cat. Data on the cat (auditory nerve) are from Kiang *et al.* (1965); data on the cochlear nucleus of the cat are from Pfeiffer (1963); superior olive data are from Tsuchitani and Boudreau (1967); and the data on the rat cochlear nucleus are from Møller, 1970. (From Møller, 1972a.)

Hz that the frequency increases. The high frequency skirts of frequency threshold curves of many of the cells in the cochlear nucleus are steeper than those of auditory nerve fibers. It is common to find nerve cells for which the slope of the high-frequency skirt of the threshold curve becomes negative, which means that the frequency has to be decreased when intensity is increased in order to retain the threshold (Møller, 1969a).

Iso-Rate and Iso-Intensity Curves

Frequency selectivity of nerve cells and fibers may be expressed in forms other than frequency threshold curves. When the intensity required for a tone to give rise to a certain discharge rate is plotted as a function of the frequency of a stimulus tone, *iso-rate* contours are obtained. *Iso-intensity curves* show spike counts at equal sound levels as a function of tone frequency.

Compared to frequency threshold curves, iso-rate curves and iso-intensity curves describe the characteristics of responses to pure tones above the threshold. (Hind, Anderson, Brugge, & Rose, 1967; Hind, Rose, Brugge, & Anderson, 1970; Rose, Hind, Anderson, & Brugge, 1971; Geisler *et al.*, 1974). In experiments designed to yield such curves, the tones are usually presented in bursts of 50–500 msec duration.

Iso-rate curves as a rule are similar in shape to frequency threshold curves. Figure 2.16 shows such curves for primary auditory fibers. Examples of iso-rate curves for a nerve cell in the cochlear nucleus are shown in Figure 2.17, along with the unit's frequency tuning curve. These curves are essentially the same shape as the frequency tuning curves. Iso-intensity curves, shown in Figure 2.18, are quantitatively different from iso-rate curves. The discrepancy between iso-intensity curves and tuning curves has not yet been explained satisfactorily.

Iso-rate and iso-intensity curves can hardly be obtained at levels of the ascending auditory pathway higher than the superior olive because they are based on a sustained train of discharges in response to tone bursts.

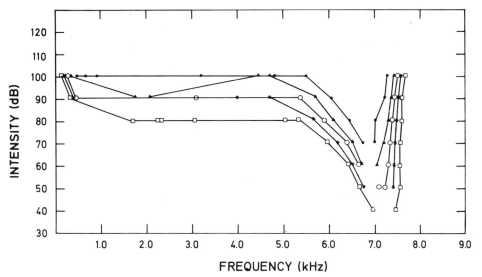

FIGURE 2.16. Iso-rate contours obtained from the auditory nerve of a squirrel monkey. (From Geisler *et al.*, 1974.)

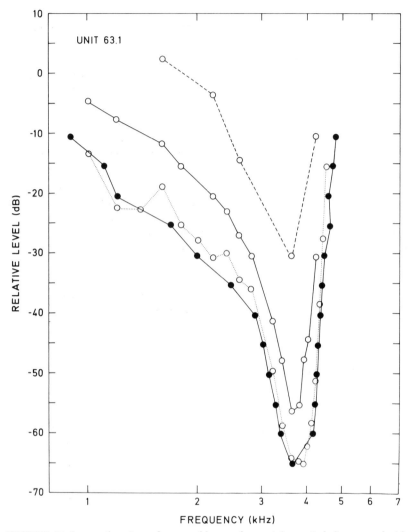

FIGURE 2.17. Iso-rate functions of a rat cochlear nucleus unit (open circles) compared with the tuning curve of the same unit (closed circles). The iso-rate curves represent the average numbers of 3.1, 4.7, and 6.2 discharges per every 50-msec-long tone burst. (From Møller, 1969a.)

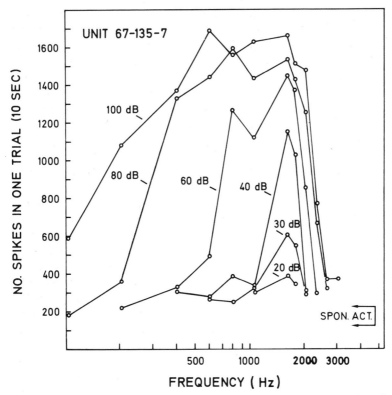

FIGURE 2.18. Iso-intensity curves from a unit in the auditory nerve of a squirrel monkey. The curves show the number of discharges evoked by stimulation by 10-sec-long pure tones of different frequencies and intensities. The sound intensity (SPL) is shown by legend numbers. (From Rose *et al.*, 1971.)

Neurons at higher levels rarely respond with a sustained train to discharges of stimulation with tone bursts.

FREQUENCY TUNING AT HIGHER LEVELS

Tuning in response to pure tones is also a prominent property of neurons at higher levels. However, since neurons at higher levels than the nuclei of the superior olive generally respond poorly to pure tones, responses are seen only at the onset or offset of a tone burst. Many units do not respond at all to pure tones. The responses of neurons at higher levels are generally more difficult to interpret, and tuning to pure tones can therefore only be determined as frequency threshold curves. However, it is clear from available results that frequency tuning in response to pure tones is prominent in single nerve cells in the nucleus of the inferior colliculus, and in the neurons of the geniculate body, as well as in primary cortical cells. The frequency threshold curves obtained at these levels of the ascending auditory pathway show a great variation in shape, particularly in width, from unit to unit, with their width varying from very narrow to very wide (Aitkin & Webster, 1972). Many cortical units are also tuned to a certain frequency in response to tone-burst stimulation, but the shape of their frequency threshold curves is more variable than that at more peripheral levels. They can be very narrow, very wide, or multiple-peaked.

It has been suggested (Katsuki, 1966) that some funneling mechanism makes tuning curves progressively narrower up to the medial geniculate (thalamic) level. Relevant experimental results, however, do not support this hypothesis. The results of experiments show that frequency threshold curves at higher levels of the auditory system may be very narrow as well as very wide (Erulkar, 1959; Rose, Greenwood, Goldberg, & Hind, 1963). In short, a great diversity prevails in the shape of the frequency threshold curves observed at these higher levels of the ascending auditory pathway, and to a greater extent in the auditory cortex than at the collicular level. In addition, neurons at higher levels respond most actively to natural, transient sounds, although many do not respond at all to tones and clicks.

Studies of the activity of single cells in the cortex are hampered by several factors. One is anesthesia, which modifies responses and renders interpretation difficult. Recordings from single cells in unanesthetized, unrestrained animals have been made possible. Such experiments yield a small amount of data compared to experiments in anesthetized animals. Furthermore, the influence of the animal's attention is difficult to measure or to control. An attempt to control attention has involved a technique combining recording of unit activity with conditioning and rewards for correct responses. However, it is difficult to determine what type of stimulus constitutes an adequate stimulus. When more types of sounds are incorporated in the battery of test signals used, the fact emerges that most nerve cells respond to sound.

Practically all studies show a considerable diversity in the response patterns of cortical units. Cells are not only specific in their response regarding frequency (spectrum). The time pattern is also of importance and there are units that only respond to sound if the sound intensity is within narrow limits. Natural sounds, in particular, such as the animal's own call sounds are efficient in evoking responses in cortical neurons (Newman & Symmes, 1979). Studies using complex sounds, including speech and speechlike sounds, suggest the existence of detectors of specific sound structures. The results presented show that specific neurons responding to different vowels may exist (Kallert, David, Finkenzeller, & Keidel, 1970; Kallert & Keidel, 1973).

It should be emphasized again that the results presented in this chapter regarding frequency tuning represent the response to pure tones, and the vast majority of the results represent the threshold (frequency threshold curves). Experimental results regarding the responses of single auditory nerve fibers and cells in the cochlear nucleus indicate that the response to complex and time-varying sounds cannot be inferred from knowledge of the threshold in response to pure tones. Results of frequency tuning studies in the form of frequency threshold curves should therefore be interpreted with caution, and it may be that the degree and type of frequency selectivity that such results represent have little functional significance.

OTHER MEASURES OF FREQUENCY SELECTIVITY

In addition to the responses of single nerve elements to pure tones, there are other manifestations of frequency selectivity. Statistical signal analysis methods have made it possible to obtain quantitative measures of the frequency selectivity of the auditory system in the presence of stimuli other than pure tones (de Boer, 1969; Møller, 1977). Grouping of the response to click sounds in accordance with the damped oscillatory response of the basilar membrane is one measure of spectral selectivity in the auditory periphery (Kaing et al., 1965a). Generally, the spectral selectivity values to which complex sounds give rise differ among themselves and collectively from those resulting from pure tones. It is thus not possible to draw conclusions about the frequency selectivity of the auditory system in response to complex sounds on the basis of knowledge about the response to simple sounds. This will be discussed in Chapter 3.

FREQUENCY SELECTIVITY WITHOUT SPECTRAL RELATIONSHIP

In the cochlear nucleus there are certain neurons in which the discharge rate is related to the *periodicity* of a sound and unrelated to the *spectrum* of a sound. When the ear is stimulated with periodic click sounds, these neurons will respond precisely to each click sound as long as the repetition

UNIT 63.6   UNIT 65.1

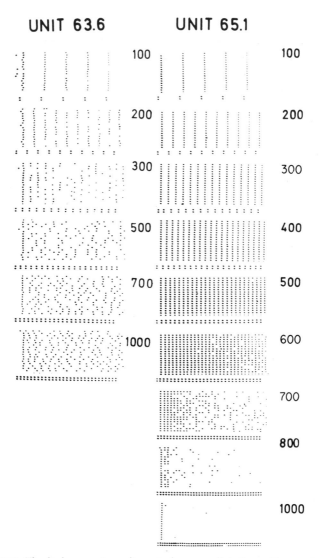

FIGURE 2.19. The discharge pattern of two single nerve cells in the cochlear nucleus of a rat in response to stimulation with repetitive clicks with various repetition rates (indicated by legend numbers in clicks per second). Each nerve impulse is indicated by a dot. The clicks were presented in 50-msec-long bursts and the time of presentation of each click is indicated by double dots below each group of responses. The left column shows the responses of a unit which belongs to the class of units that respond to pure tones with a sustained train of discharges. The right column shows the responses of a transient unit. (From Møller, 1969b.)

rate remains below a certain value (Møller, 1969b). If the intervals between the click sounds are reduced below a certain value, the units abruptly cease firing (Figure 2.19). Plotting the discharge rate as a function of click frequency illustrates how such a unit is "tuned" to a certain click rate (Figure 2.20). It becomes evident that this selectivity is not based on the spectrum of the click train when it is shown that the response does not change when the polarity of every second click is reversed, whereupon the spectrum is changed. It becomes evident that these units are not general periodicity detectors when it is found that these units only respond to transient sounds, such as clicks and the onset of tone bursts with rapid rise time. Also, other experiments have shown that it is not the periodicity, as such, that these nerve cells selectively respond to but rather it is the duration of silences between sounds that determines this reponsiveness. That has been shown in experiments where short tone bursts have been used as stimuli rather than clicks. Results of these experiments showed that when

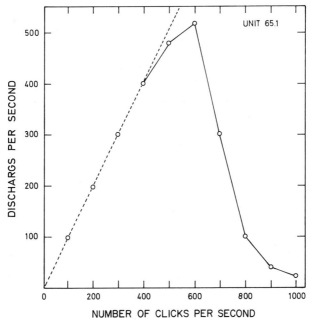

FIGURE 2.20. Illustration of periodicity tuning in a transient-type neuron in the cochlear nucleus. The discharge rate is shown as a function of click repetition rate. The dashed line indicates one discharge per click. (From Møller, 1969b.)

these units once have fired an action potential, they require a certain period of silence before they can fire again (Figure 2.21). This time period varies from unit to unit and is not an effect of the refractory period but probably of a delayed inhibition that switches the unit off immediately after it has fired an impulse. As long as this inhibition is in effect, the unit cannot refire (Møller, 1969b).

Inhibition

TWO-TONE INHIBITION

Two-tone inhibition is closely related to frequency tuning. This phenomenon can be demonstrated in recordings from single auditory nerve fibers when two tones are presented simultaneously. By proper choice of frequency and intensity of the two tones, it can be shown that one tone reduces the discharge rate evoked by the other tone.

Activity evoked by a tone at CF is most easily suppressed by a tone in the frequency range just above the fiber's response range. Moreover, in most cases, tones below the CF can also inhibit the discharges evoked by a tone at CF (Nomoto, Suga, & Katsuki, 1964; Sachs & Kiang, 1968; Sachs, 1969; Arthur, Pfeiffer & Suga, 1971). Just as frequency threshold curves may be plotted to describe the threshold of a fiber for one tone, it is also possible to construct tuning curves to show the threshold of the tone that suppresses the activity evoked by another tone at the fiber's CF (Figure 2.22).

Although the physiological mechanism behind suppression is not yet known, two-tone inhibition has been shown to occur with the same delay as does excitation (Arthur et al., 1971). In this light, suppression cannot result from inhibition in the usual neurophysiological sense (i.e., involving an interneuron), since this would cause a delay of at least .5 msec in the inhibition. Theories have advanced that the second tone may reduce the deflection (vibration) of the basilar membrane evoked by the first tone in a purely hydromechanical manner. Excitation of hair cells on both sides of the maximally exicited ones would thereby be reduced (Legoux, Re-

---

FIGURE 2.21. *Upper graph:* Similar graph as in Figure 2.20 comparing responses to repetitive tone bursts and clicks. The straight line indicates one nerve impulse per tone burst or click. *Lower graph:* The same data as in the upper graph but replotted to show the average discharges per sound (tone burst or click). The symbols are the same as in the upper graph. (From Møller, 1969b.)

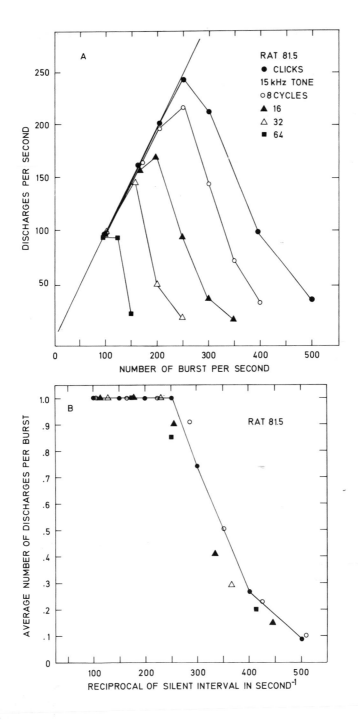

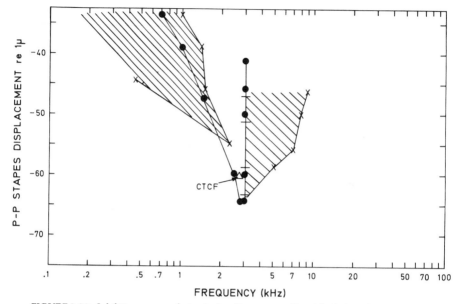

FIGURE 2.22. Inhibitory areas of a typical auditory nerve fiber (shaded area) together with the frequency threshold curve (filled circles). The suppression areas were determined in terms of the threshold for inhibition of the activity evoked by a tone at a frequency and intensity marked CTCF. (From Nomoto *et al.*, 1964.)

mond, & Greenbaum, 1973). It has been shown in experiments, in which the intracellular potentials in inner hair cells have been recorded by microelectrodes, that two-tone inhibition is present at the level of the inner hair cells (Sellick & Russell, 1979). These results strongly support the previously mentioned hypothesis that two-tone inhibition is closely related to the vibration of the basilar membrane itself. In addition, there is experimental evidence that two-tone inhibition is closely related to the nonlinearity of the inner ear. These results also provide strong evidence against the hypothesis that two-tone inhibition is a result of interaction between adjacent hair cells, (Kim, Siegel, & Molnar, 1979; Kim, Molnar, & Matthews, 1980; Rhode, 1977).

As to the functional importance of inhibition, it was assumed that inhibition facilitated frequency resolution in the auditory system, and thus was similar to lateral inhibition that was shown to exist in the eye. (The role of the latter is believed to be enhanced spatial contrast.) As more and more knowledge about two-tone inhibition is accumulated from neurophysiology experiments, its role of increasing spectral resolution has become doubtful. It may be questioned whether inhibition has anything at all to do with spectral resolution. It now seems more likely that it has to do with processing of time-varying sounds (see Chapter 4).

In many cochlear nucleus units, inhibition similar to two-tone inhibition in primary auditory fibers can be seen. Thus, the discharge rate evoked by a tone at CF can be reduced or suppressed in most cochlear nucleus units by a second tone of a certain frequency and intensity. Generally, a tone having a frequency about 20% higher than the CF of the unit is most efficient for this purpose, but in many cases tones having a frequency below the unit's CF can be inhibitory (Rose et al., 1959; Greenwood & Maruyama, 1965; Møller, 1969c, 1971, 1976a, b). The two-tone inhibition in most nerve cells in the cochlear nucleus is thus very similar to that in primary auditory nerve fibers. It might, therefore, be hypothesized that two-tone inhibition occurs at the input to these neurons and that the two-tone inhibition occurs at the input to these neurons and that the cochlear nucleus nerve cells only relay information carried by primary auditory nerve fibers. That hypothesis is supported by the fact that there is no discernible time delay between inhibition and excitation in most of the cells. That also rules out the possiblity that there could be an interneuron present in the inhibitory path (Møller, 1976a). Contrary to this hypothesis and contrary to the situation in fibers in the cochlear nerve, is the fact that the spontaneous activity of many nerve cells in the cochlear nucleus is also inhibited. Another equally plausible hypothesis is that neurons in the cochlear necleus receive input from many primary nerve fibers, some of which terminate on inhibitory synapses and some of which terminate on excitatory synapses. This is supported by the finding that inhibition seen in the cochlear nucleus may be affected by drugs that are applied locally to a small region of the cochlear nucleus from which the recording was made. Thus, Watanabe (1979) found that the drug picrotoxin has a "disinhibitory" effect on the firing of single nerve cells in the cochlear nucleus in that when applied locally, picrotoxin could reversibly abolish the inhibitory effect of one tone on the activity evoked by another tone.

Whereas the response pattern of nerve fibers in the cochlear nucleus is relatively homogeneous, neurons in the cochlear nucleus exhibit a variety of response patterns. One group of neurons, which have a primary-like response pattern with regard to two-tone inhibition, is represented by those discussed earlier. There are also neurons with other patterns of response. Thus, some few neurons have inhibition that occurs with a longer delay than does excitation. Neurons of this group are likely to have another type of two-tone inhibition than that found in primary nerve fibers. The longer latency of the inhibition indicates that an interneuron is included in the inhibitory pathway. Furthermore, a few units in the cochlear nucleus, particularly in the dorsal part, have a high spontaneous activity and cannot be excited by tones at all. However, their discharge frequencies can be reduced or inhibited by a tone within a certain frequency-intensity range.

OTHER TYPES OF INHIBITION

Whereas two-tone inhibition is the only type of inhibition seen in the responses of primary auditory nerve fibers, different and more complex types of inhibition are also seen—not only at higher levels of the ascending auditory pathway—but in the cochlear nucleus, where some cells show an interplay between excitation and inhibition that is different for different sound intensities, and where the time delays between excitation and inhibition are also often different (Møller, 1976a). Some cells in the cochlear nucleus can be inhibited by contralateral stimulation (Mast, 1970).

One complication in experiments concerned with inhibition is related to the effect of anesthesia. While anesthesia does not seem to influence responses from cells in the anterior and posterior ventral cochlear nucleus greatly, it has been shown to alter the responses from cells in the dorsal cochlear nucleus (Evans & Nelson, 1973) to a great extent. Interplay between inhibition and excitation in these latter neurons is strongly influenced by anesthesia. Anesthesia seems to affect inhibition more than excitation, and its effect increases as one moves centrally along the ascending auditory pathway. The inhibitory and excitatory input, which is received by many of these higher-order neurons from other parts of the brain, is also likely to be modified by anesthesia. Results obtained in anesthetized animals may therefore not be representative of the response in the awake, alert animal, particularly regarding the interplay between inhibition and excitation.

The response pattern to sounds that change with time and binaural sounds cannot be predicted on the basis of results obtained using steady state sounds. It is possible that the main importance of the interplay between inhibition and excitation, which is seen consistently within the entire auditory system, has its greatest importance in connection with time-varying sounds.

Tonotopic Organization

Throughout the ascending auditory pathway, nerve cells and fibers are arranged anatomically in accordance with their characteristic frequencies. In the cat, the auditory nerve fibers with low CF are located in the center of the nerve bundle (Kiang *et al.*, 1965a). Within each of the three major subdivisions of the cochlear nucleus (AVCN, PVCN, and DCN), neurons are arranged according to CF from high to low frequencies in the dorsal-ventral, posterior-anterior, and medial-lateral directions. This tonotopic organization is a result of the orderly branching and distribution of auditory nerve fibers (Figure 2.23; Rose *et al.*, 1959). This means that a microelectrode that passes through each of these subdivisions will record from cells with CF which might be arranged from high to low frequencies. This means that individual auditory nerve fibers innervate relatively restricted areas of the brain nuclei, from which the information passes on toward the auditory cortex.

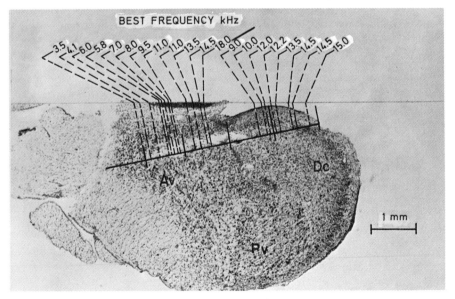

FIGURE 2.23. Tonotopic organization in the cochlear nucleus of the cat. The CFs of the nerve cells encountered in the electrode track illustrated are given by legend numbers. Av is anterior ventral nucleus; Dc, dorsal nucleus; and Pv, posterior ventral nucleus. (From Rose *et al.*, 1959.)

Similar organization is found in other brain nuclei. In the superior olivary nucleus, tonotopic organization has been shown by Goldberg and Brown (1969) and by Tsuchitani and Boudreau (1966). The lateral lemniscus also has a tonotopically organized distribution of auditory nerve cells as described by Aitkin, Anderson and Brugge (1970). The inferior colliculus (Rose *et al.*, 1963) and the medial geniculate (Aitken & Webster, 1972) also are tonotopically organized (see also Clopton, Winfield & Flammino, 1974). The tonotopic organization in the auditory cortex was demonstrated early in the history of auditory electrophysiology by Woolsey and Tunturi (see Woolsey & Walzl, 1942; Tunturi, 1950).

Research by Merzenich and Brugge and their colleagues (Merzenich, Knight, & Roth, 1975; Middlebrooks *et al.*, 1979; Merzenich, Anderson, & Middlebrooks, 1980; Merzenich & Brugge, 1973; Imig & Adrian, 1978; Imig & Brugge, 1978; Colwell & Merzenich, in press) has shed new light on the functional organization of the auditory cortex and have shown that the cochlea is represented on the surface of the cortex in a more complex way than was speculated from ealier work. Extensive microelectrode mapping in the cat revealed that there are four separate fields that are organized according to frequency (see Figure 2.24). In this figure, AI, AAF, PAF, and VPAF comprise four principal cortical divisions of the cochleotopic auditory system. AII and the temporal fields (cross-hatched in the figure) comprise the cortical fields of a parallel diffuse auditory system. The AI area exhibits a highly ordered representation of the basilar membrane of the cochlea with the apical part of low frequencies located rostrally and high frequencies caudally. The older data are largely consistent with the newer data when differences in experimental methods are considered.

The older data were obtained by recording compound action potentials using gross electrodes, while the more recent data were obtained using microelectrodes. Different animal species show different auditory cortex representations, and the exact localization of the auditory area differs from animal to animal within the same species. This tonotopic organization may have to do with neural integration, causing it to occur preferentially between nerve cells that are tuned to a similar frequency.

It has been shown that neurons in the cat's cortex are organized in stripes according to their binaural response properties. These bands are orthogonal to the iso-frequency contours (Merzenich *et al.*, 1977; Middlebrooks *et al.*, 1980; Imig & Adrian, 1978; Imig & Brugge 1978; see Figure 2.25). In this cortical field the bands that are perpendicular to the iso-frequency contours (dashed lines in Figure 2.25) have predominantly

AUDITORY CORTICAL FIELDS

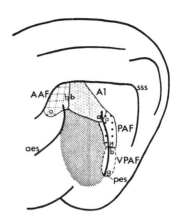

EARLIER VIEW
(after Woolsey)

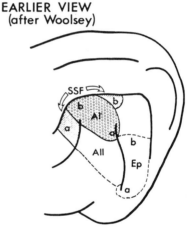

FIGURE 2.24. Schematic representation of the basic organization of auditory cortical fields in the cat. AI = primary auditory field; AAF = anterior auditory field; PAF = posterior auditory field; VPAF = ventroposterior auditory field. The region of representation of the cochlear apex (a) and cochlear (b) is indicated for all four of these cochleotopically organized fields. (From Merzenich et al., 1980.)

binaurally excitatory–excitatory or excitatory–inhibitory (shaded areas) responses. There are also "non-tonotopic" auditory cortical fields in which the different neurons have very broad tunings. The best frequencies of these neurons are therefore difficult to determine, but when it is possible to do so there is no apparent orderly representation of units with different best frequencies. In the cat, there are several such fields

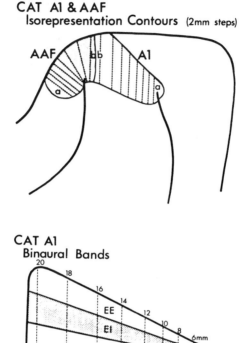

FIGURE 2.25. Internal organization of the cochleotopically organized auditory fields in the cat. *Top:* Representation of the cochlea within two auditory fields, the "primary" (AI) and the anterior field (AAF). The dotted lines show equal representation of the cochlear sensory epithelium about 2 mm apart. The first of these lines represent a point on the basilar membrane of the cochlea that is 6 mm from the extreme apex. *Bottom:* Schematic representation of the binaural bands of primary auditory cortex (A) in the cat. Contours representing points along the cochlear sensory epithelium 2 mm apart (6–20 mm from the apex) are illustrated. (From Merzenich *et al.*, 1980.)

with a diffuse frequency representation (Merzenich, 1979) and at least one such field has been found in primates (Merzenich & Kaas, 1980).

Previously, attention was focused on the organization of neurons according to their frequency tuning in response to pure tones. Recently, interest has been directed to tuning to other properties of sounds. The most extensive work has been performed by Suga (1981), who showed in the bat's cortex there are areas where neurons are specifically tuned to such properties of the bat's echolocation sounds as are related to ascertaining distance to the reflecting object (ranging). Specifically, Suga found three different specialized functional areas designed to process: (a) different types of echolocating sounds such as Doppler-shifted echoes (speed of object); (b) a combination of the frequency modulation of the orientation sounds and specific harmonics of the reflected sound; and (c) responses to the continuous part of the echolocation sounds and their harmonics.

Thus, it seems that cells of the same animal species may give rise to selective responses in certain neurons. Similar results have been obtained in the squirrel monkey where it has been shown that certain time segments of the call sound give rise to responses in specific neurons and not in others (Newman & Symmes, 1979). The functional importance of this finding is undetermined, but work done particularly in the bat (Suga, Kuzirai & O'Neill, 1981), and in the squirrel monkey (Newman & Symmes, 1979), seems to indicate that there may be neurons in these cortical areas that act as "feature detectors" that are "tuned" to more complex and more functional features of a sound than just its frequency.

These studies have concerned the natural sounds of these various animals and only a few species and a few sounds have been studied. It is not known how these results can be extrapolated to humans in whom natural sounds are speech sounds. Specific areas for the different components of normal speech may exist, but it is a task of the future to explore them.

Discharge Rate as a Function of Stimulus Intensity (Rate Intensity Curves)

It has generally been assumed that the discharge rate in a certain fiber of the auditory nerve is a measure of the level of excitation of one or a few hair cells. It has also been assumed that excitation of a single hair cell represents the energy in a certain frequency band due to the frequency selectivity of the basilar membrane. Thus, the discharge rate in a certain nerve fiber is assumed to carry information about the energy level in a certain frequency band to higher nervous centers. This is the basis of the place hypothesis of frequency discrimination. It is therefore of interest to consider the relationship between stimulus intensity and discharge rate in single auditory nerve fibers.

Above threshold, the discharge rate of primary fibers in response to noise or pure tones increases monotonically with stimulus intensity up to a certain level (Figure 2.26). Most auditory nerve fibers display a certain low discharge rate, even in a silent environment. The range of sound intensities between threshold and saturation, however, is only about 30 dB (Kiang *et al.*, 1965a).

Stimulus response functions of single auditory nerve fibers to tone bursts generally have a sigmoidal shape, but the slope of the curves depends on whether the tone is below or above the CF of the nerve fibers (Figure 2.27; Sachs & Abbas, 1974). For tones above the CF, the slope is

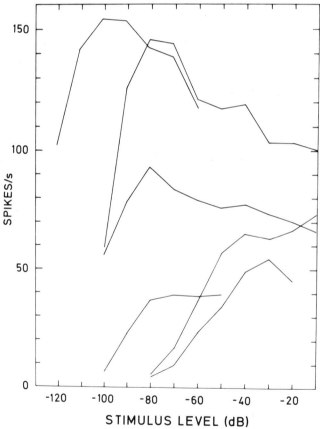

FIGURE 2.26. Typical stimulus response curves of single auditory nerve fibers. The graph shows the discharge rate (spikes per second) as a function of relative level of the stimulus tones. These were continuous tones with a frequency equal to the fiber's CF. (From Kiang *et al.*, 1965a.)

less and the maximal discharge rate has a much lower value than for tones below. Whether this is related to the strong inhibitory response areas at frequencies above the CF or to nonlinearity in the motion of the basilar membrane has yet to be resolved.

The discharge rate of most cochlear nucleus units increases with increasing sound intensity up to a certain level, beyond which it becomes constant. This is known as saturation. In the cochlear nucleus, many cells have stimulus response curves resembling those of primary auditory nerve fibers (Figure 2.28; Møller, 1969a). Other units display a non-monotonic stimulus response relationship, for example, their discharge rate reaches a peak at a certain sound intensity, above which it decreases. In some units, discharge rate diminishes to nearly zero at a certain high sound intensity. It should be emphasized that these response characteristics were seen under rather unnatural circumstances inasmuch as the stimulus consisted of pure tones of constant frequency and amplitude.

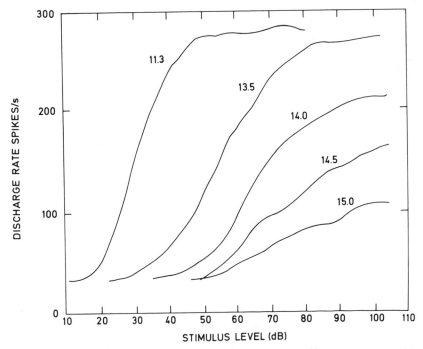

FIGURE 2.27. Stimulus response curves of a single auditory nerve fiber in response to tones of different frequencies (indicated by legend numbers) at and above the fiber's CF (11.3 kHz). (From Sachs & Abbas, 1974.)

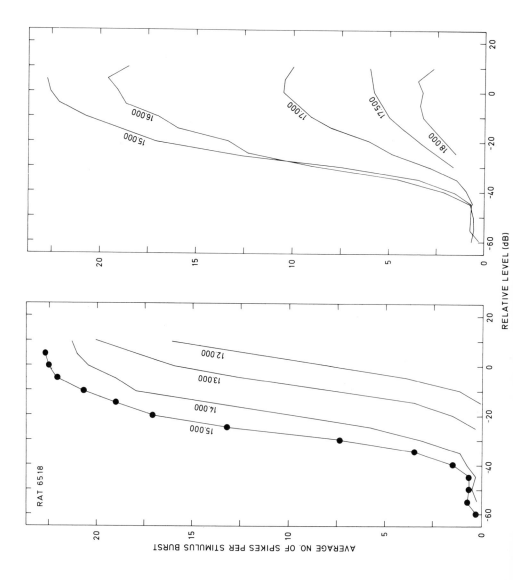

148

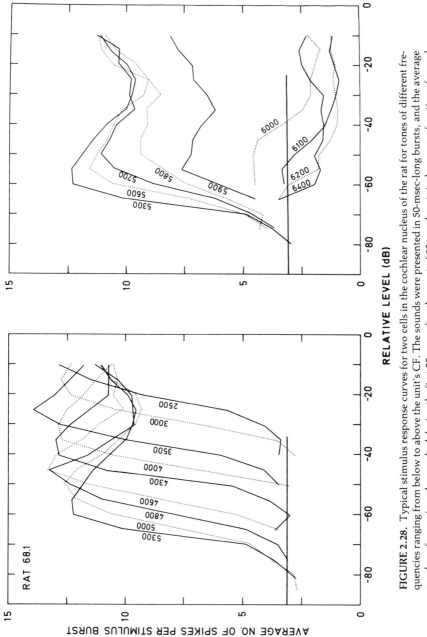

FIGURE 2.28. Typical stimulus response curves for two cells in the cochlear nucleus of the rat for tones of different frequencies ranging from below to above the unit's CF. The sounds were presented in 50-msec-long bursts, and the average number of nerve impulses evoked during the first 55 msec after the onset of 32 tone bursts is shown as a function of sound intensity. The two left hand graphs show responses to tones below CF and the right hand graphs represent tones above CF. The CF of the two units was 15 kHz (upper graphs) and 5.3 kHz (lower graphs). (From Møller, 1969a.)

149

Stimulus intensity curves of responses of neurons in the superior olive complex have a sigmoid shape (Boudrea & Tsuchitani, 1970). In these neurons as well, the maximal obtainable discharge rate is lower for tones above CF than at or below CF. They are similar in this respect to fibers of the auditory nerve and the cochlear nucleus.

It is also important to consider that signaling of sound intensity in the ascending auditory pathway obviously cannot be performed by any one single fiber because the intensity range between threshold and saturation of the discharge rate is smaller than the range of intensities over which the auditory system can function. Signaling of sound intensity to higher nervous centers must therefore be achieved through the interaction of many fibers. Activities of these fibers may then be integrated at some higher center in the brain. It has been assumed that there were different groups of fibers with different threshold. Such an assumption offers a simple explanation of extending intensity by integrating the neural activity of many fibers. However, extensive recordings from single nerve fibers show that fibers within a certain range of CF's have similar thresholds (Liberman & Kiang, 1978; Kiang et al., 1965a). Signaling of sound intensity may therefore be more complex in nature than originally assumed.

Assumptions about the nature of the coding of sound intensity rest on results obtained using pure tones. Studies of the responses of both primary auditory nerve fibers and neurons in the cochlear nucleus to more complex sounds, which were nonetheless steady, yielded different results. Results of such studies may, in fact, indicate that the discharge rate of auditory nerve fibers is kept within narrow limits by some feedback mechanism (Rose et al., 1971, Møller, 1977). This is based on the finding that the average discharge rate in response to steady sounds is relatively independent of the stimulus intensity for levels some 20–30 dB above threshold, whereas the waveform of the sound is reproduced in the temporal pattern of the discharges to about the same degree throughout a wide range of intensities (more than 80 dB). This is called phase-locking and is thus restricted to sounds with frequencies below 4–5 kHz. When dealing with high frequency sounds in units with a high CF, the situation may be slightly different. This will be discussed in detail in Chapter 3.

Temporal Coding of Frequency of Pure Tones

The previous section concerned the spectral selectivity of the peripheral auditory system and the spectral or place coding in the discharge pattern of nerve fibers and nerve cells in the ascending auditory pathway. This section deals with temporal coding of frequency. This principle of frequency coding is based on the finding that the discharge pattern of auditory nerve fibers, as well as many cells in the various nuclei of the peripheral part of the ascending auditory pathway, is phase-locked to the periodicity of low-frequency sounds. Phase-locking of the neural discharge to the periodicity of a sound was first reported by Galambos and Davis (1943) and is defined as when the neural discharges tend to group together at a certain phase of a sine wave or other periodic sounds. Phase-locking is only seen for low-frequency tones—below about 5-6 kHz in single auditory nerve fibers.

The tendency to discharge synchronously with the periodicity of a sine wave was previously thought to reflect a more or less perfect locking of discharges to certain phase of the sound wave. Contemporary views, emphasizing the stochastic rather than the deterministic properties, however, hold that the *probability* of firing varies systematically over a period of the stimulus and is thereby always greatest at a certain phase of the stimulus waveforms. Distinct phase-locking does occur, but not as commonly as was once assumed.

Phase-locking also exists in the discharges of many nerve cells in the cochlear nucleus, superior olive, inferior colliculus, and medial geniculate, as these show a certain degree of synchrony or phase-locking with the waveform of the sound when the ear is stimulated by low frequency pure tones. The upper frequency limit of phase-locking is lower in the cochlear nucleus than in primary auditory nerve fibers, and it is lower again in the neurons of the inferior colliculus.

Interdependence of neural discharges and the temporal pattern of sound, particularly of a pure tone, has been shown in recordings from single auditory nerve fibers using various methods (Tasaki, 1954; Rupert, Moushegian, & Galambos, 1963; Nomoto et al., 1964; Hind et al., 1967; Rose, Gross, Geisler, & Hind, 1966; Brugge, Anderson, Hind, & Rose, 1969a; Anderson, Rose, Hind, & Brugge, 1971; Sachs et al., 1974.) The simplest method is to display nerve impulses on an oscilloscope, the sweep of which is locked to a fixed point on the stimulus waveform, and then to observe whether the nerve impulses are spread evenly over the period of the sound wave or are clustered around a certain phase of the sound wave (Tasaki, 1954; see Figure 2.29). This is a *qualitative* method. A common *quantitative* method is to use period histograms that are locked to the stimulus sound wave and show the distribution of nerve impulses over one stimulus period (Kiang et al., 1965a; Rose, Brugge, Anderson, & Hind, 1967). When data are collected over a sufficient period of time, such period histograms provide an estimate of the prob-

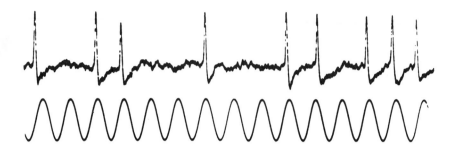

FIGURE 2.29. Synchronization of responses of a cochlear fiber to low-frequency tones illustrated by a continuous film record of the response of a guinea pig fiber to a tone with a frequency of .3 kHz and a low intensity (near threshold). (From Arthur *et al.*, 1971.)

ability of firing at various points of time during the period of the stimulus sound and a quantitative index of phase-locking can then be obtained. In single nerve fibers of the auditory nerve, modulation of the discharge pattern, according to the sound's waveform, gradually decreases above a certain frequency, due to the temporal smoothing inherent in the neural excitatory process, and it is hardly measurable at frequencies above 5–6 kHz.

Sounds of a greater complexity than pure tone sounds produce period histograms in shapes that are nearly the same as the half-wave rectified version of the waveform of the sound for frequencies up to about 5–6 kHz, as shown in Figure 2.30 (Rose *et al.*, 1971). Such histograms thus depict the excitation of the hair cells.

It has been demonstrated that the time pattern of a low-frequency sound modulates the discharge pattern of individual auditory nerve fibers, even when the sound is within an intensity range above within which the mean discharge rate reaches saturation. This means that the auditory nerve fibers can continue transmitting information about rapid changes, despite saturation. In this light, the probability of firing of an individual nerve fiber is a precise function of the half-wave rectified time

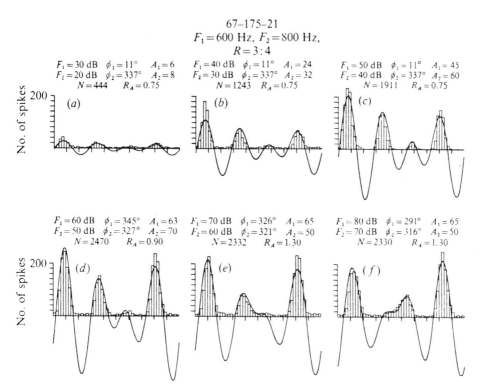

67-175-21
$F_1 = 600$ Hz, $F_2 = 800$ Hz,
$R = 3:4$

FIGURE 2.30. Period histograms of the activity recorded from a fiber in the auditory nerve of the squirrel monkey in response to stimulation with a complex periodic sound composed of two sinusoidal tones locked together with a frequency ratio of 3 : 4. The amplitude ratio between the two tones was kept constant at 10 dB while the overall intensity was varied over a 50 dB range. The amplitude and phase of the two tones used to construct the fitted waveforms are specified in each graph. (From Rose *et al.*, 1971.)

pattern of the sound up to a certain frequency (Rose *et al.*, 1971; Møller, 1977).

Phase-locking that is prominent in all fibers of the auditory nerve in response to low-frequency pure tones, or other periodic sound, gradually disappears as information passes the neurons of the brainstem nuclei of the ascending auditory pathway. The upper frequency, where phase-locking can be shown, decreases toward higher levels of the ascending auditory pathway. Many, but not all, of the neurons of the cochlear nucleus show a prominent phase-locking similar to that which occurs in the auditory nerve (Galambos & Davis, 1943; Rupert & Moushegian,

1970; Lavine, 1971; Caspary, 1972; Goldberg & Brownell, 1973; Rose *et al.*, 1967; Britt, 1976). In the superior olive nuclei, the discharge patterns of many neurons are still phase-locked to low-frequency sounds (Moushegian, Rupert, & Whitcomb, 1964; Goldberg & Brown, 1969; Brownell, 1975), whereas in the inferior colliculus only a few of the neurons have phase-locked responses to low-frequency tones (Rose, Gross, Geisler, & Hind, 1966). In the medial geniculate nucleus, almost no neurons show phase-locked responses (Rouiller, de Ribaupierre, & de Ribaupierre, 1979), and, in the primary auditory cortex, no phase-locking has been observed in single unit responses.

Nonlinearities in the Auditory Periphery

Nonlinearity in the transduction of the hair cells that originates from their sensitivity to deflection in only one direction (unidirectionality) is accepted by all, but that is not the nonlinearity considered here. The nonlinearity considered here regards the generation of combination tones. When two or more tones are presented simultaneously a nonlinear relationship exists between the excitation level and the response (saturation nonlinearity). Nonlinearity can be demonstrated by psychoacoustic experiments as well as by neurophysiological experiments. For many years the middle ear was considered to be the source of the nonlinearity, but that theory is no longer accepted.

It may be convenient to divide nonlinearity into two types—one that gives rise to nonlinear distortion, and one that is slow and does not produce nonlinear distortion products. Slow saturation nonlinearity may best be compared to an automatic gain control mechanism in that it does not produce distortion products or intermodulation products but makes the input–output amplitude relationship become nonlinear.

Although it is well documented from psychoacoustic experiments that under many circumstances the ear exhibits various types of nonlinear phenomena, there has been a great deal of controversy about how the observed phenomena arise and where in the ear the sources of these nonlinearities reside. Until recently, the physiological methods used to study nonlinearity utilized recordings of the cochlear microphonic potential (see Wever & Lawrence, 1954). Recordings from single nerve fibers in experimental animals have contributed significantly to understanding the physiological mechanisms that underlie the generation of nonlinear distortions in the ear. Experimental results now available clearly indicate that the site of the nonlinearity is the inner ear. It is still strongly debated however whether it is the motion of the basilar membrane that is nonlinear and the source of this nonlinearity or whether nonlinearity is associated with the properties of hair cells.

Nonlinearities of the ear have been studied in neurophysiological experiments (Nomoto *et al.*, 1964; Goldstein & Kiang, 1968; Goblick & Pfeiffer, 1969; Anderson *et al.*, 1971; Abbas & Sachs, 1976; Smoorenburg, Gibson, Kitzes, & Rose, 1976; Greenwood, Merzenich & Roth, 1976; Russell & Sellick, 1978; Kim, Siegel & Molnar, 1979; Sellick & Russell, 1980). In these experiments the cubic difference tone $(2f_1 - 2f_2)$ that appeared when two tones were presented simultaneously with a small frequency difference was found to be the most prominent distortion product, but the existence of other distortion products has also been demonstrated. Thus, the discharge pattern of single auditory nerve fibers may be shown to be phase-locked to $2f_1 - f_2$ and $f_2 - f_1$. This has been demonstrated from period histograms of the responses of single auditory nerve fibers in response to the two tones shown in Figure 2.31 (Goldstein 1967, 1970; Goldstein & Kiang, 1968). When histograms are locked to either of the two primary tones, they naturally show periodic patterns (modulations), indicating that a large number of the discharges are phase-locked to either of the two primary tones. However, when these histograms are locked to the frequency of the cubic difference tone $(2f_1 - f_2)$, the histogram appears to be substantially modulated (Figure 2.31).

Whether the $2f_1 - f_2$ component is generated in the cochlea and whether or not a mechanical disturbance exists in the inner ear (on the basilar membrane) that corresponds to that component have been matters of discussion and of much experimental effort. Many of the results of neurophysiological experiments, as well as the results of psychoacoustic experiments, seem to indicate that the $2f_1 - f_2$ component is subjected to the same spectral filtering in the inner ear as is a sound that is transmitted to the cochlea in the ordinary way. This supports the hypothesis that distortion product exists as a mechanical disturbance in the cochlea.

Attempts by Rhode (1977) to detect cubic difference tones in the vibration of the basilar membrane were, however, generally unsuccessful, at least at sound levels similar to those used in the psychoacoustic as well as in the neurophysiological experiments. The amplitude of the cubic difference tone that could be observed in the cochlea was, in general, much lower than that observed psychoacoustically. These results, then, speak against the hypothesis that components exist as mechanical events in the cochlea.

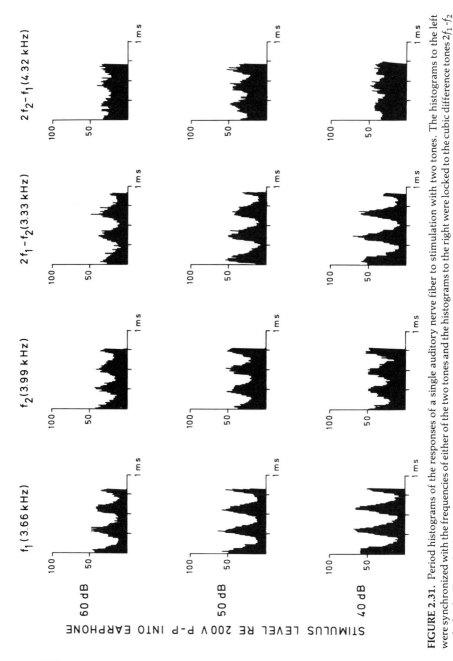

FIGURE 2.31. Period histograms of the responses of a single auditory nerve fiber to stimulation with two tones. The histograms to the left were synchronized with the frequencies of either of the two tones and the histograms to the right were locked to the cubic difference tones $2f_1 - f_2$ and $2f_2 - f_1$. (From Goldstein & Kiang, 1968.)

156

Kim et al., (1979), however, find strong evidence that distortion products $2f_1 - f_2$ are generated on the basilar membrane—where the primaries
are at highest amplitudes—and that these distortion products are prop-
agated along the basilar membrane as a disturbance much like any other
real sound. Dallos (1977) showed that, in a cochlea in which the outer hair
cells were missing for the region responding with the highest amplitude to
f_1 and/or f_2, $2f_1 - f_2$ were not generated. His results were confirmed by
Siegel (1978), who showed that the distortion products, $f_2 - f_1$ and $2f_1 -$
f_2, which could be detected in the discharge pattern of single auditory
nerve fibers in the chinchilla, were altered significantly when those parts
of the basilar membrane that responded to the primary tones were damaged by noise exposure. Compared to the situation in the ears of normal
animals, which showed large distortion products under the same conditions, in damaged ears, no measurable distortion products could be
detected. Distortion products thus seem to be produced by a mechanism
that is vulnerable to anoxia and overstimulation. In cochleas where there
were reversible effects produced by exposure to tones of a lower intensity,
the amplitudes of the previously mentioned distortion products were
reduced by a factor of 2 to 3. The amplitude of the distortion products
returned to normal within a few minutes. The generation of $2f_1 - f_2$ is
thus vulnerable to both permanent (irreversible) and temporary trauma
to hair cells in the regions of the primary tones (Kim et al., 1979; Smoorenburg, 1980; Kim, 1980). This latter characteristic of the $2f_1 - f_2$ component is the most convincing evidence that nonlinearity is physiologically vulnerable.

Results of these studies indicate that distortion products are produced
at the location of maximal stimulation of the primary tones and that the
nonlinearity responsible for production of these distortion products is
physiologically vulnerable (Kim, 1980).

Distortion products corresponding to $2f_1 - f_2$ and $f_2 - f_1$ may be
recorded in physical sound in the ear canal in an experimental animal that
is fitted with a suitable earphone (Kemp, 1979; Kim et al., 1980). This
observation lends further support to the assumption that distortion products are generated in the ear. By making use of the observation that distortion products are vulnerable to anoxia and cyanide, some researchers
(Kim, 1980; Siegel et al., 1982) presented evidence that their origin is in
the cochlea and not in the middle ear, or in the sound-generating system.
The cochlea in these experiments was perfused with KCN solution. After
the experiment, the neural response (N_1, N_2) disappeared and the distortion products in the ear canal also disappeared (Kim, 1980).

Thus, the nonlinearity responsible for the generation of $2f_1 - f_2$ components seems to be dependent upon hair cell integrity in the region where the primaries give maximal excitation. Reduction in function (e.g., fatigue) of the hair cells in the high-frequency part of the cochlea also reduces $2f_1 - f_2$, even though the sensory epithelium, at the frequency of $2f_1 - f_2$, is intact. Even reversible fatigue at frequencies of the primaries reduces the amplitude of the combination tones.

It is likely, therefore, that the nonlinearity in the cochlea is related to the function of the hair cells. It may be suggested that as a result of overstimulation the changes in the mechanical properties of the hairs may change the mechanical properties of the basilar membrane and be responsible for a change in production of difference tones. It is possible that the hairs contribute a variable or nonlinear coupling between the basilar membrane and the tectorial membrane. Findings that show that the hairs of the hair cells contain actin may eventually explain the nonlinearity (Flock & Cheung, 1977; Flock, 1980). It was shown earlier (p. 122) that the frequency selectivities of single auditory nerve fibers are physiologically vulnerable. Also, certain physiologically vulnerable phenomena are involved in the generation of the cochlear microphonics (Pierson & Møller, 1980a, b). The latter indicates that there is nonlinear interaction between two generators for the cochlear microphonic potential. Such an interaction may explain the phase shift that is seen with intensity in the generation of $2f_1 - f_2$.

There are several important discrepancies between psychophysical and neural data. Thus, the phase angle of the $2f_1 - f_2$ combination tone decreases 6–12° for each decibel increase in stimulus intensity, whereas the neurophysiologically measured phase angle of this distortion product is nearly independent of stimulus amplitude (Goldstein, 1967; Goldstein & Kiang, 1968; Goldstein, 1970; Furst & Goldstein, 1980). The psychoacoustically measured combination tone $2f_1 - f_2$ increases nonmonotonically with increase in the intensity of the lower primary tone (Helle, 1969; Smoorenburg, 1972). In addition psychoacoustic data show some unexplained jumps in phase angle of as much as 360° as a function of sound intensity (Zwicker, 1980).

Electrophysiological studies on coding of two-tone complexes in single neurons in the cochlear nucleus show a somewhat different pattern from that seen in primary auditory fibers. Thus, Smoorenburg et al. (1976) found that both the $f_1 - f_2$ and the $2f_1 - f_2$ components increase approximately in proportion to increases in the intensity of the primary tones.

Response to Complex Sounds: Time-Varying Sounds

ADAPTATION

The simplest time-varying sound is the tone burst. Tone bursts used in studies of the auditory system usually consist of a tone that has a relatively fast onset, then has a constant amplitude for a period of time, and then is switched off so that its amplitude decreases rapidly to values below the threshold of the ear. Responses from single nerve fibers in the auditory nerve and from nerve cells in the cochlear nucleus to such time-varying sounds have been studied in great detail. Results are usually shown in the form of peristimulus time (PST) histograms. These histograms show the average distribution of discharges after the onset of the tone burst, as obtained when responses to a large number of tone bursts are added. PST histograms of responses from single auditory nerve fibers show that these fibers respond during the entire duration of the tone burst, but the peak seen in the PST histogram a few milliseconds after the onset of the tone indicates that more nerve impulses are evoked in the beginning of the tone burst than later during the steady-state part of the tone. The decrease in height of the histograms during the time the tone is on is understood in classical neurophysiology as a measure of the adaptation of the system (receptor and nerve fiber) under test. The decay usually has an exponential shape, and the rate of decay has been used as a quantitative measure of adaptation. Similar responses obtained from other sensory systems and receptors have been used as a basis for classifying these systems as tonic (nonadapting), slowly adapting, or fast-adapting systems, depending on if and how fast the response decays after its initial value. Adaptation is actually an indication that the system responds more or less preferentially to changes in the intensity of a stimulus. It is usually inferred that a fast-adapting receptor or nerve cell responds better to changes in stimulus intensity than to steady-state sounds. Using this classification system, it is inferred that responses of primary auditory fibers show little to moderate adaptation, whereas neurons in the cochlear nucleus show a varying degree of adaptation.

Adaptation saves channel capacity in the nervous system. A continuous sound, once it starts to repeat itself, ceases to bear information. Certainly, the longer a tone is observed, the more accurately its frequency can be determined. However, when the time and frequency resolution of the ear reach their limits, not much is gained by further analysis. Like most other sensory systems, nerve cells within the auditory system adapt to continuous stimulation. For example, the discharge rate of auditory nerve fibers in response to tone bursts is high in the beginning but decreases to a value that is maintained as long as the tone bursts continue (Kiang *et al.*, 1965a). The degree of adaptation is different at various levels of the ascending auditory nervous system and is usually more pronounced at central levels compared to peripheral levels. In fact, most neurons in the cortex respond poorly to steady-state sounds.

Figure 2.32 shows typical recordings from a single auditory nerve fiber in response to tone bursts. Whereas each tone burst is seen to evoke discharges, the number of discharges varies from one tone burst to another, illustrating some of the statistical properties of neural excitation.

Due to the statistical variation in the discharge frequency of most auditory neurons, it is necessary to average the obtained responses for many stimulus presentations. Peristimulus time (PST) histograms[2] of the recorded discharges, which show the mean response (spike-density) to many consecutive stimulus presentations, are commonly used for such averaging. When a sufficient amount of data is averaged, the histograms are a valid approximation of the probability of firing during one such stimulus of the sound. Examples of such histograms are shown in Figure 2.33. These nerve fibers respond throughout the duration of the tone, but the number of discharges decreases with time after the onset. That in-

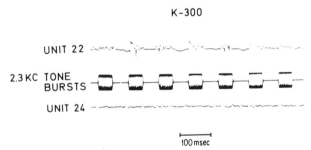

FIGURE 2.32. Discharges from a single nerve fiber in the auditory nerve in a cat in response to tone bursts. (From Kiang *et al.*, 1965a.)

---

[2] These are often referred to as *poststimulus time histograms*, but it seems more correct to use the word *peristimulus time histograms*, since these histograms usually show the distribution of nerve discharges during a time period beginning at the onset of the sound and not beginning at the end of the stimulus. This becomes important when the stimulus is a long tone burst.

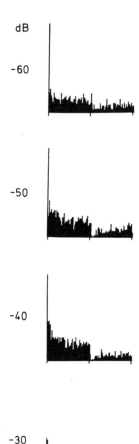

dB

-60

-50

-40

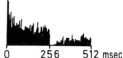

-30

0   256   512 msec

FIGURE 2.33. Typical peristimulus-time histograms
of the responses of an auditory nerve fiber in a cat.
(From Kiang *et al.*, 1965a.)

dicates that primary auditory nerve fibers only have a moderate degree of
adaptation.

In contrast to the responses from neurons at higher levels (including the
cochlear nucleus), which show a great variety of different shapes of PST
histograms in response to tonebursts, the PST histograms of responses of
auditory nerve fibers to tonebursts show little variation among different
fibers. Generally, examination of the decay of such PST histograms
shows that adaptation becomes more and more pronounced as one gets

higher in the ascending auditory nervous system. Some neurons do not show a simple exponential decay in the PST histogram, but the PST histograms of the response from higher neurons usually have a more complex shape.

In the cochlear nucleus there are many nerve cells that show little to moderate adaptation, but there are also nerve cells that show a high degree of adaptation. In the inferior colliculus, only a few nerve cells show a tonic response to tone bursts. Neurons in more centrally located nuclei, such as the medial geniculate and the neurons of the auditory cortex, generally respond only to the onset or offset of a tone burst. The cochlear nucleus has units that respond to tone bursts in the same general way as do primary fibers, but the discharge rate of the cochlear nucleus units decreases more rapidly with time, indicating a larger degree of adaptation than occurs with primary fibers. Some units in the cochlear nucleus respond only to the onset of the burst, indicating a large degree of adaptation (Kiang et al., 1973; Kiang et al., 1965a, b; Evans & Nelson, 1973; Godfrey, Kiang, & Norris, 1975a; Godfrey, Kiang, & Norris, 1975b).

On the basis of the shape of the PST histograms (Pfeiffer, 1966), four different classes of neurons have been distinguished in the cochlear nucleus (Figure 2.34). The histogram shown in Figure 2.34A has a similar shape to those of the responses from primary auditory nerve fibers with a low degree of adaptation. The one in Figure 2.34B has a typical damped oscillation (chopper type), and the histogram in Figure 2.34C represents another type with a characteristic pause following the initial response; this type was named "pauser." The histogram in Figure 2.34D represents a type of neuron that responds only to the onset of a tone burst. This last class of neuron is interesting, since neurons that belong to that class respond selectively to the repetition rate of repetitive transients (click sounds), independent of the spectrum of the sound. The response pattern of this type of neuron is described in detail on p. 133.

Intracellular recordings from nerve cells in the cochlear nucleus have provided further information about the synaptic interaction in the cochlear nucleus (Britt & Starr, 1976; Gerstein, Butler, & Erulkar, 1968; Starr & Britt, 1970). Responses of the chopper type of neurons were studied by intracellular recordings by Romand (1979).

Superior olive neurons often show prominent adaptation, although it is not uncommon for them to respond to tone bursts with a sustained train of impulses, thus reflecting little or moderate adaptation. In the superior olive, PST histograms of the units' responses to tone bursts have a variety

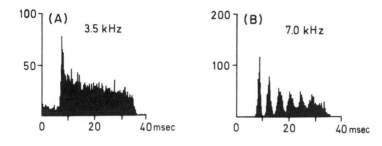

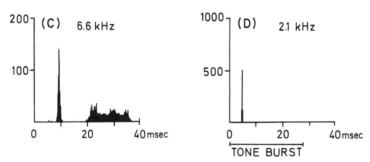

FIGURE 2.34. PST histograms of responses of cells in the cochlear nucleus to tone bursts. Each histogram represents one class of units. (From Pfeiffer, 1966.)

of shapes. Many units have response patterns resembling those of cochlear nucleus units (Goldberg, Adrian, & Smith, 1964; Tsuchitani & Boudreau, 1966). Prominent adaptation is most characteristic at high sound intensities.

In the lateral lemniscus, complex patterns of response, in which the shape of the PST histograms changes with intensity, are seen. The various neurons of the inferior colliculus, medial geniculate, and auditory cortex, which tend to respond only when the stimulus changes, do so in a more pronounced manner as the stimulus intensity increases. Some neurons, particularly those in the auditory cortex, will also respond to the offset of a tone burst. The temporal firing pattern of medial geniculate neurons is rather complex, and the PST histograms of their responses to pure tones have many different shapes with multiple peaks spaced evenly and unevenly (Figure 2.35).[3] Generally, both the inferior colliculus and the medial geniculate exhibit a wide variety of response patterns when simple sounds such as clicks and pure tones are used as stimuli. Onset response, offset response, and periodic responses are common, whereas sustained responses are rare. The overall picture is one of strong excitatory-inhibitory reactions (Aitkin, Dunlop, & Webster, 1966). In light of the rapid adaptation in the auditory system, it may be surprising that the sensation of a tone burst does not fade away after some time.

When evaluating the response to tone bursts, it should be kept in mind that these sounds are artificial sounds. Tone bursts are sounds that are switched on abruptly, usually from below threshold, to an intensity that is held constant until the tone is switched off. Natural sounds seldom reach levels below threshold, and there is usually a background sound. Normally, the intensity of natural sounds gradually varies around a certain value, and the variations are often relatively small. Since the auditory system cannot be regarded as functioning as a linear system functions, one cannot generalize about the responses to time-varying sounds on the basis of responses to tone bursts. Experiments using different types of amplitude-modulated sounds have shown that clearly. It is therefore important to use sounds that are more similar to natural sounds

---

[3] A word of caution similar to that expressed for coding of the temporal pattern of sound may be appropriate when interpreting PST histograms. Here as well, we must rely heavily on temporal averaging of the responses to the same stimulus repeated a number of times in order to reduce the statistical variation inherent in the responses to a single stimulus. The nervous system itself, in contrast, carries out a *spatial* averaging, by virtue of which the responses of many fibers are averaged. This decreases the statistical variability of the firing of the discharges of a single nerve cell to a single stimulus presentation.

In addition to that, the stimuli used in such experiments are usually highly artificial; namely, tone bursts that rise rapidly in intensity from below threshold attain a constant value for a certain time, and, in the end, decrease rapidly to below threshold. No natural sound has these characteristics. In units having no spontaneous activity and a relatively constant discharge frequency such a stimulus, when repeated, will evoke the first discharges at very much the same point in time after the onset of the stimulus. Due to the relatively periodic firing rate of these units, even the second discharge is likely to occur within about the same period of time after the first discharge and then the third discharge, and so on. The PST histograms may then acquire the shape of a damped oscillation, as exemplified by the "chopper" type seen in Figure 2.33. The averaged responses of a *population* of neurons are not likely to have the same pattern since the mean discharge rates of the individual neurons in the population differ slightly. The discharges of many nerve cells may not be synchronized. This is an example in which *averaging* over time will give different results from *averaging over many nerve fibers* (spatial averaging). In many cases, the temporal averaging may be a valid approximation of the spatially averaged response of a number of fibers having equal probabilities of firing, but this, by no means, may be taken for granted.

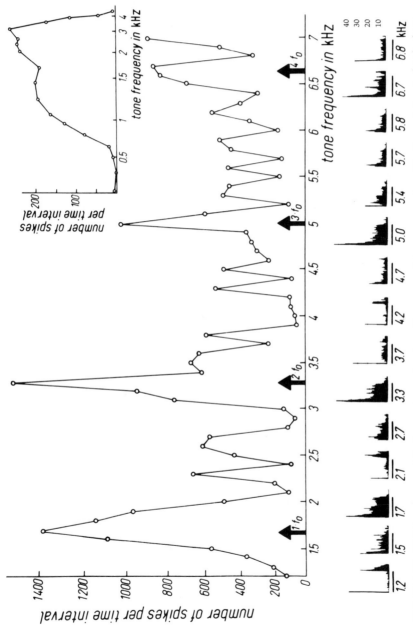

FIGURE 2.35. PST histograms and discharge rate as a function of frequency of the responses of a cell in the medial geniculate body in response to tone bursts. (From Kallert, 1974.)

165

than tone bursts and clicks in experiments investigating signal transformation in the auditory nervous system. Sounds such as amplitude- and frequency-modulated tones and noise are examples of sounds for which the rate of change, as well as the degree of change in amplitude and frequency, can be easily varied and the characteristics of such sounds are well defined.

RESPONSES TO AMPLITUDE- AND FREQUENCY-MODULATED SOUNDS

Tones or noises that are amplitude or frequency modulated with sinusoids or other waveforms are more akin to natural sounds. However, very few studies have been reported in which such sounds have been used. When a low degree of modulation is maintained, the dynamic properties of the system can often be studied as though the system were functioning linearly. At this stage, a few examples of response pattern to amplitude- and frequency-modulated tones will be given.

AMPLITUDE-MODULATED TONES

Discharges of neurons in the cochlear nucleus are phase-locked to the modulation when the ear is stimulated with amplitude-modulated tones, provided that the modulation frequency is kept within certain limits (Glattke, 1969; Møller, 1972b, 1974a, b). Examples of such phase-locking are illustrated in Figure 2.36, which shows period histograms of the responses to a tone that is amplitude modulated with a sinusoid. The two histograms are the responses to tones modulated with 25 and 200 Hz, respectively. If compiled over a sufficient length of time, histograms reveal the probability of firing at various phases of the modulation cycle. As shown in Figure 2.36, histograms are modulated to a different degree for different modulation frequencies. The relationship between the modulation of the histograms and the modulation frequency of the stimulus sound is shown in more detail in Figure 2.37. When relative amplitude of modulation of the period histograms is plotted as a function of the modulation frequency, a modulation transfer function is obtained. This may be displayed in a *Bode plot* showing the ratio in decibels between modulation of the histogram and modulation of the sound as a function of modulation frequency. A Bode plot of the modulation transfer function often includes the phase angle between modulation of the sound and modulation of the histogram (shown as a function of modulation frequency). This graph in Figure 2.37 shows the modulation

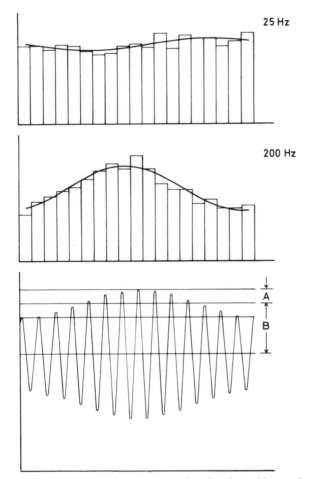

FIGURE 2.36. Period histograms of the responses of a cell in the cochlear nucleus of a rat to an amplitude-modulated tone. The frequency of the tone that was modulated was equal to the cell's CF (15 kHz) and results from two different modulation frequencies (25 and 200 Hz) are shown. One period of the modulated sound is shown in the histograms. (From Møller, 1974b.)

transfer function obtained from a cell in the cochlear nucleus. From this example, it clearly follows that the modulation of histograms is greater for a certain range of modulation frequencies. In that range, a few decibels of amplitude modulation of the stimulus sound can modulate the discharge rate more than 50%, provided that the modulation frequency is

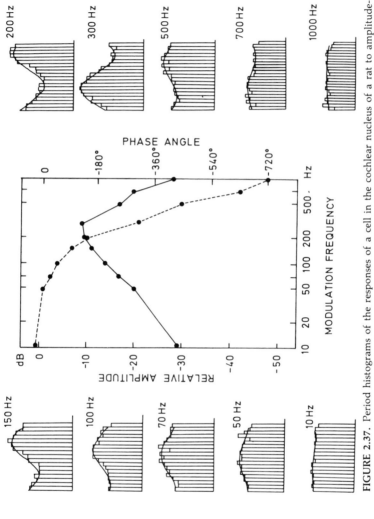

FIGURE 2.37. Period histograms of the responses of a cell in the cochlear nucleus of a rat to amplitude-modulated tones for modulation frequencies between 10 and 1000 Hz. The middle graph is the modulation transfer function showing the relative amplitude of the modulation of the histograms as a function of the modulation frequency. The dashed line gives the phase angle between the modulation of the histogram and that of the stimulus sound. The frequency of the tone was equal to the unit's CF. (From Møller, 1972b.)

within a certain range. The frequency at which the greatest modulation will occur varies from unit to unit, but the greatest enhancement is usually around 100–500 Hz (Møller, 1972b). The relative degree of modulation may change with sound intensity—usually the modulation is attenuated more at low than at high frequencies when the sound intensity is increased. Usually, modulation frequencies between 50 and 500 Hz are preserved over a broad range of sound intensities. This range extends far beyond the range of sound intensities, where the discharge rate in response to steady sounds has reached a plateau (Møller, 1974a, b). As shown in Figures 2.36 and 2.37, modulation of the histograms is closely sinusoidal. The sinusoidal waveform of the modulation is thus reproduced almost undistorted in the firing pattern.

The conclusion that may be drawn from the results shown in Figure 2.36 and 2.37 is that small and rapid changes in the intensity of a sound are reproduced to a great extent in the discharge rates of these units. These neurons may signal small changes in sound intensity to higher centers, even though they are unable to signal the mean sound intensity.

It is of interest to note that in the presence of an unmodulated tone with frequency equal to the unit's CF, amplitude modulation of an inhibitory tone produces a modulation of the discharge rate in a similar way as modulation of a single tone at the unit's CF (Møller, 1975a). (As mentioned earlier, nerve cells in the cochlear nucleus as well as in the fibers of the auditory nerve have inhibitory areas that surround the excitatory area. These inhibitory areas are most pronounced in the frequency range above the excitatory area of the unit.) As shown in Figure 2.38, modulations of the cycle histograms are, however, reversed in phase compared with what is the case when a tone at CF is amplitude modulated. When a tone at CF is modulated and an unmodulated tone with a frequency that is equal to the best inhibitory frequency of the unit is added, the range of intensities over which the modulation of the tone at CF is reproduced in the discharge pattern is extended (Møller, 1975a). These results show that small changes in the amplitude of a tone are reproduced in the discharge pattern of cochlear nucleus units to a greater extent when another (unmodulated) sound is present.

Further studies using amplitude modulated sounds showed that the change in the amplitude of an inhibitory tone is reproduced in the discharge pattern with exactly the same latency as are changes in the amplitude of an excitatory tone (Møller, 1975b). These studies show that either the inhibitory and excitatory inflow to these neurons (inhibitory and excitatory synapses) are on the same nerve cell or that the inhibition

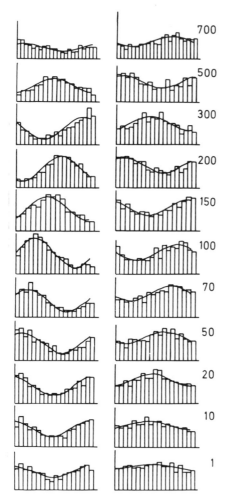

FIGURE 2.38. Period histograms of the responses to stimulation with two tones, one of which was amplitude modulated. Left: Modulation of the excitatory tone; Right: Modulation of the inhibitory tone. The frequency of the excitatory tone was equal to the unit's CF (4500 Hz) and its intensity was 55 dB SPL. The frequency of the inhibitory tone was equal to the unit's best inhibitory frequency (5500 Hz) and its intensity was 65 dB SPL. The frequency of the modulation is given by legend numbers. (From Møller, 1975a.)

that is present in the response of cochlear nucleus units actually takes place in the cochlea and is the same as suppression (see p. 136).

Only a few studies have been concerned with respose patterns of single neurons in the inferior colliculus to frequency- and amplitude-modulated sounds. Results reported by Erulkar, Nelson, and Bryan (1968) show a complex pattern, with, in general, a great enhancement of small changes in both frequency and intensity of a pure tone (Figure 2.39).

TONES WITH CHANGING FREQUENCY

The response to tones the frequencies of which were changed over a large range, has been studied in relatively few experiments. Those that have been studied have been restricted, for the most part, to neurons in the cochlear nucleus and in the nucleus of the inferior colliculus.

Response patterns of cochlear nucleus units to tones whose frequencies vary over a large range and at a slow rate (sweep tones) resemble iso-intensity curves (see p. 128). When the frequencies of such tones are changed at a rapid rate, the response pattern is greatly altered. The pattern is, in fact, a function of the rapidity of the change in stimulus frequency (Møller, 1971). This is shown in Figure 2.40 which shows that the range of frequencies over which responses are obtained is narrower when the frequency of the tone is varied rapidly. The total number of nerve impulses (per unit time) is almost the same in these two situations. Variation in the rate of frequency change, then, causes a redistribution of the nerve impulses so that more discharges are evoked near the unit's CF using rapid as compared to slow variation of stimulus frequency (Møller, 1971, 1974b).

As the rate of change in frequency is increased, the height of the peaks in histograms of responses from cochlear nucleus units increases to a certain rate of change in frequency, above which there is a rapid decrease in peak height. There is thus an optimal rate of change in frequency at which nerve cells respond with the most localized distribution of nerve impulses. This optimal rate of change varies from nerve cell to nerve cell and, in the rat, it is most commonly in the range of 1–15 MHz/sec (Møller 1969c, 1971, 1972a).

It may be worth noting that the responses to tones and noise of rapidly varying frequencies are maintained over a large frequency range, and unlike many features of constant sounds, the response pattern to changing sounds is maintained—and even enhanced—over the physiological range of sound intensities. Background masking noise affects the response only to a small extent. Primary auditory nerve fibers do not show the same degree of change in response pattern to tones with high rate of change in frequency as do cochlear nucleus units (Sinex & Geisler, 1981). It may therefore be assumed that enhancement of the response to sounds

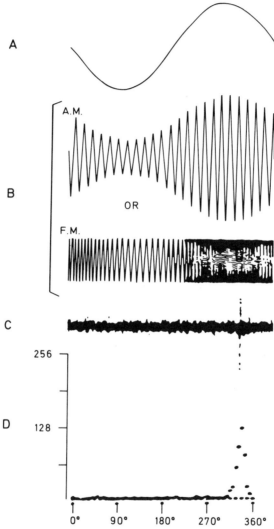

FIGURE 2.39. Responses of a cell in the inferior colliculus to amplitude-modulated tones. The modulation was sinusoidal and the tone was modulated to 50% at different rates given by legend numbers. (From Erulkar *et al.*, 1968.)

UNIT 120.12

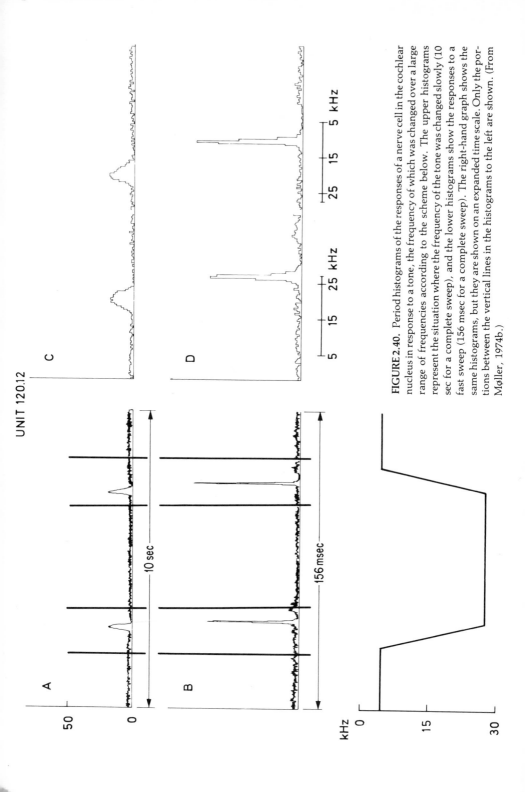

FIGURE 2.40. Period histograms of the responses of a nerve cell in the cochlear nucleus in response to a tone, the frequency of which was changed over a large range of frequencies according to the scheme below. The upper histograms represent the situation where the frequency of the tone was changed slowly (10 sec for a complete sweep), and the lower histograms show the responses to a fast sweep (156 msec for a complete sweep). The right-hand graph shows the same histograms, but they are shown on an expanded time scale. Only the portions between the vertical lines in the histograms to the left are shown. (From Møller, 1974b.)

with rapid change in frequency (spectrum) that is seen in the cochlear nucleus is a result of processing of information in the nervous system (cochlear nucleus) and not a property of the inner ear.

Binaural Interaction in the Ascending Auditory Pathway

Localization of a sound source in the horizontal plane is likely to involve several binaural cues. Low-frequency sinusoids and transient sounds are assumed to be discriminated on the basis of their interaural time or phase differences, whereas high-frequency sounds are discriminated on the basis of their interaural intensity differences. Both processes require a neural connection between the left and right ascending auditory neural pathways. Furthermore, it is likely that discrimination, using the interaural time difference entails comparison of the neural activity at a level of the ascending auditory pathway where the temporal characteristics are still preserved in the discharge pattern. A connection would thus have to exist at a relatively peripheral location on the auditory pathway. The nucleus of the superior olive complex is known to be the first site of connections from both ears.

Neurons of the medial superior olive (MSO) have been shown to respond to stimulation from both ears. The responses of many MSO neurons are affected by stimulation of either ear, but usually in different ways. Some neurons receive excitatory input from either ear (excitatory–excitatory: EE), and other neurons receive excitatory input from one ear and inhibitory from the other (excitatory–inhibitory: EI). In most of these latter cases, the contralateral input is excitatory and the ipsilateral is inhibitory. In the LSO part of the nucleus, ipsilateral stimulation is usually excitatory, whereas contralateral input is inhibitory. The frequency tuning of those neurons that respond to both ipsilateral and contralateral stimuli is usually similar for the two ears.

Response patterns of many units in the superior olive nucleus complex are direct functions of a binaural time difference (phase difference) on stimulation with transient sounds or low-frequency sinusoids (Goldberg & Brown, 1969). Many neurons in the superior olivary complex respond with lowest threshold in response to binaural clicks when there is a certain time delay between the clicks (Hall, 1965). Above threshold, the highest probability of response occurs at the same time delay where the threshold is lowest. The EI units are sensitive to interaural intensity differences (Goldberg & Brown, 1969). Some units do not respond when the sound intensities in the two ears are the same, but do respond when there is a difference in the sound intensity in the two ears. Similar findings have been obtained in the inferior colliculus (Hind, Goldberg, Greenwood, & Rose, 1963; Rose, Gross, Geisler, & Hind, 1966; Geisler, Rhode, & Hazelton, 1969; Beneveto & Coleman, 1970) and the medial geniculate (Aitkin & Webster, 1972). Brugge, Anderson, and Aitkin (1970) found that similar binaural interaction occurs in neurons in the dorsal nucleus of the lateral lemniscus (for a review see Erulkar, 1975). In the auditory cortex, Brugge *et al.* (1969b) found a similar sensitivity to interaural time difference. Examples of their findings are shown in Figure 2.41 (for a review, see Goldberg, 1975). Merzenich and Kaas (1980) found that there was, in addition to the tonotopic organization, specific binaural organization of the auditory cortex. Thus, responses to binaural sounds were organized on the cortex according to the time difference between the sound reaching the two ears. This topical representation of difference between the sound in the two ears is likely to be of great functional importance.

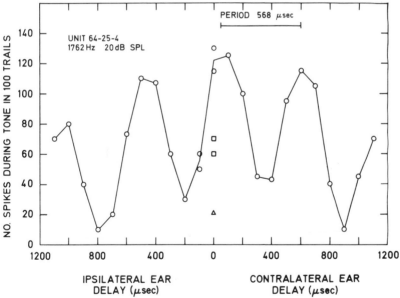

FIGURE 2.41. Discharges evoked by a nerve cell in the cortex in response to clicks, one click presented to each ear with a certain delay. The graph shows the discharge rate as a function of the delay (in microseconds) with the ipsilateral ear leading to the left and the contralateral ear leading to the right. (From Brugge *et al.*, 1969b.)

Previously reported results were all obtained using two sound receivers, one connected to each ear. Natural binaural stimulation, in which the interaural time difference results from a different travel time for the sound to the two ears, evokes a response of units in the inferior colliculus which bears a specific relationship to the direction of movement. Nonetheless, the majority of the units responded only when movement was within a restricted range (Altman, 1968).

The cochlear nucleus possesses neurons that receive input from the contralateral as well as the ipsilateral ear, but the functional importance of this connection is unknown. It may concern directional hearing, but it is assumed that the nervous centers of importance for directional hearing are located at higher levels of the ascending auditory pathways (superior olivary complex and the inferior colliculus). The contralateral input to neurons in the cochlear nucleus can be inhibitory or excitatory in different units. Inputs of both types have tuning curves of about the same shape as those of other neurons. The best frequency is nearly the same for both types of input.

Descending Auditory Nervous System

Parallel to the ascending auditory pathway there is a system of descending pathways, the extension of which, as we know it, is shown in Figure 2.42 (Harrison & Howe, 1975). Two descending systems emerge from the primary auditory cortex: the corticothalamic system (Figure 2.42A) and the corticocochlear system (Figure 2.42B). The first part terminates in the medial geniculate body, whereas the other system has a widely distributed network of connections to all the nuclei of the principal ascending auditory pathway, and it extends to the cochlear hair cells. The latter part of that system, called the olivocochlear system, is the best-known part of the descending system, both anatomically and physiologically. The behavioral function of the descending system is not known but it is generally believed that it has mainly an inhibitory or habituating influence on the activity in the ascending auditory pathway (Whitfield, 1967). Only a few physiological studies have been aimed at investigating its role in the processing of sound (for a review, see Klinke & Galley, 1974). It is also worth noting that the results just mentioned were obtained in nonprimate species and that a difference exists between primates and nonprimates with regard to the size of the LSO that is poorly developed in primates.

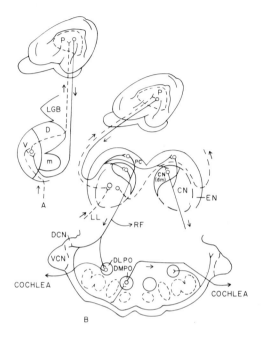

(A) THE CORTICOTHALAMIC SYSTEM

(B) THE CORTICOCOCHLEAR SYSTEM

FIGURE 2.42. Schematic outline of the descending (solid lines) and ascending (dashed lines) connections in the auditory system of the cat: (A) the corticothalamic system; (B) the corticocochlear system. *Abbreviations:* P, principal area of the auditory cortex; LGB, lateral geniculate body; D, dorsal division of medial geniculate; V, ventral division of medial geniculate; M, medial (magnocellular) division of the medial geniculate body; PC, pericentral nucleus of the inferior colliculus; CN dm, dorsal medial part of the central nucleus of the inferior colliculus; CN, central nucleus of inferior colliculus; EN, external nucleus of inferior colliculus; LL, lateral lemniscus; RF, reticular formation; DCN, dorsal cochlear nucleus; VCN, ventral cochlear nucleus, DLPO, dorsolateral periolivary nucleus; DMPO, dorsomedial periolivary nucleus. (From Harrison & Howe, 1974.)

OLIVOCOCHLEAR BUNDLE

Anatomy

The olivocochlear bundle originates in the superior olive region of the brainstem and runs in or close to the vestibular part of the auditory nerve (Spoendlin, 1966; Warr, 1978). The fibers of this bundle reach the hair cells where ear fiber bifurcates, so that each descending fiber connects with many hair cells. The total olivocochlear bundle consists of approximately 800 fibers. These fibers give rise to at least 4000 efferent endings which should be compared to 30,000–40,000 afferent endings (see Spoendlin, 1966; Tamar, 1972). It is predominantly outer hair cells in the basal part of the cochlea that have many efferent endings. They have 6 to 10 efferent endings each, compared to 5 to 8 afferent endings on each basally located outer hair cell. In the mammalian species studied, the efferent fibers that reach each cochlea come from both sides (i.e., each cochlea receives bilateral efferent input). The crossed fibers most likely originate from cells located dorsomedially in the accessory superior olive nucleus, whereas the uncrossed fibers are the processes of cells in the S segment of the superior olive nucleus (Whitfield, 1967).

Physiology

When electrically stimulated, the fibers of the olivocochlear bundle reduce both spontaneous and sound-driven activity in single afferent auditory nerve fibers (Desmedt, 1962; Fex, 1962, 1965). The effect of electrical stimulation is weaker on the uncrossed efferent bundle than on the crossed bundle. The afferent activity evoked by high-frequency tones is usually inhibited by efferent stimulation more than that evoked by low-frequency tones (Wiederhold & Peake, 1966; Wiederhold, 1970; Wiederhold & Kiang, 1970).

The cochlear microphonic potential recorded at the round window increases in amplitude on electrical stimulation of the olivocochlear bundle. It has been shown that electrical stimulation of the olivocochlear bundle alters the endocochlear potential (Fex, 1967), and other studies indicate that stimulation of the olivocochlear bundle changes the mechanical properties of the cochlea or basilar membrane (Mountain, 1980). Inhibition of afferent activity by stimulation of the olivocochlear bundle has a long latency (15–75 msec). The control of the afferent activity that these fibers produce is thus relatively slow, and the inhibition amounts, at most, to 14 dB (Desmedt, 1962).

Recordings from single efferent fibers show that the efferent fibers of the olivocochlear bundle respond to sound stimulation. These efferent fibers have regular spontaneous activity, unlike afferent fibers that have irregular spontaneous activity. Spontaneous activity of the uncrossed efferent fibers can be inhibited by stimulating the contralateral ear with sound. The crossed fibers do not seem to have that property.

Sectioning of the olivocochlear bundle has been shown to increase the sensitivity of the ear in measurements of the acoustic middle ear reflex (Borg, 1971). This indicates that the olivocochlear bundle not only acts as a negative feedback system but also exerts a tonic inhibitory action on the afferent activity. A few psychophysical experiments have been carried out to reveal what effect a disruption of the olivocochlear bundle has on sound perception. The results indicate that frequency discrimination may deteriorate slightly when the olivocochlear bundle is not functioning properly.

The efferent fibers are connected to the hair cells via a synapse, and it is therefore assumed that the activity in the efferent fibers affects the sensitivity of the hair cells via a transmitter substance. It is, however, not yet clear what the transmitter substance is, although several studies indicate that the nerve endings of the crossed olivocochlear bundle are cholinergic (Guth, Norris, & Bobbin, 1976; Bobbin & Thompson, 1979). These findings were supported by those of Comis and Leng (1979), who showed that infusion of acetylcholine inhibited reversibly the size of the compound action potential (N_1) near threshold while it was ineffective at levels well above threshold. It is somewhat puzzling however, that the effect of the olivocochlear bundle can be abolished by strychnine. That would indicate that the transmitter substance is glycine.

OTHER DESCENDING SYSTEMS

As shown in Figure 2.42, all nuclei of the ascending auditory pathway receive input from higher levels. Without much experimental evidence at hand, it is assumed that this input is mainly inhibitory in nature, but a few studies indicate that stimulation of a descending fiber tract can increase the discharge rate in the receiving nucleus.

Electrical stimulation of the auditory (AI) has been shown to inhibit some of the activity in the cells of the medial geniculate body. Such experiments indicate a negative feedback between the primary cortex and the medial geniculate which may be frequency specific. Likewise, cells of the auditory cortex may send fibers to the inferior colliculus (Diamond, Jones, & Powe, 1969), to the dorsal nucleus of the lateral lemniscus, and to regions as far down as the superior olive and trapezoidal body. Also, there are connections between the AII area of the auditory cortex and the medial geniculate body (Diamond et al., 1969; Winer, Diamond, & Raczkowsky, 1977) and similar efferent connections exist between the AAF area and the medial geniculate (Diamond et al., 1969).

There is also anatomical evidence for the existence of efferent connections from the medial geniculate body to the inferior colliculus (van Noort, 1969), but little is known about the physiology of this efferent pathway. There are two efferent bundles in the lateral lemniscus that terminate in the dorsal cochlear nucleus and the periolivary nuclei. The lateral nucleus of the trapezoid body and the lateral superior olivary nucleus send axons to the ipsilateral cochlear nucleus (AVCN) (Warr, 1969). The descending system, no doubt, receives information not only from other auditory centers in the brain but also from the brainstem reticular formation and the cerebellum. Of particular interest is a connection between the cerebellum and the inferior colliculus shown by Snider and Stowell (1944; see also Aitkin & Boyd, 1978).

Response of cells in the cochlear nucleus can often be inhibited by contralateral sound stimulation (Mast, 1970). Whether this is the result of a descending pathway is not known, but the long latency would suggest that such is the case. It is not known if the descending system plays a role in binaural hearing and specifically in sound localization, but the crossing and bilateral innervation that are present in the descending system may indicate such a role (Harrison & Howe, 1975).

Information about the function of the auditory system in this chapter rests, in most cases, on relatively well-established experimental results. Such experimental results are not available for all parts of the auditory system, and there are parts of the system about which we know very little. The transformations of complex sounds in higher brain nuclei are examples of important parts of the auditory system, about the function of which we know very little. The role of the descending nervous system and that of adrenergic (autonomic) influence on the inner ear are other such examples.

References

Abbas, P. J., & Sachs, M. B. Two tone suppression in auditory nerve fibers. Extension of a stimulus response relationship. *Journal of the Acoustical Society of America*, 1976, *59*, 112–122.

Aitkin, L. M., & Boyd J. Acoustic input to the lateral pontine nuclei. *Hearing Research*, 1978, *1*, 67—477.

Aitkin, L. M., Dunlop C. W., & Webster, W. R. Click-evoked response patterns of single units in the medial geniculate body of the cat. *Journal of Neurophysiology*. 1966, *29*, 109–123.

Aitkin, L. M., Anderson, D. J., & Brugge, J. F. Tonotopic organization and discharge characteristics of single neurons in nuclei of the lateral lemniscus of the cat. *Journal of Neurophysiology* 1970, *33*, 421–440.

Aitkin, L. M., & Webster, W. R. Medial geniculate body of the cat: Organization and responses to tonal stimuli in neurons in ventral division. *Journal of Neurophysiology,* 1972, *35*, 365–380.

Altman, J. A. Are there neurons detecting direction of sound source notion? *Experimental Neurology*, 1968, *22*, 13–25.

Anderson, D. J., Rose, J. E., Hind, J. E., & Brugge, J. F. Temporal position of discharges in single auditory nerve fibers within the cycle of a sine-wave stimulus: Frequency and intensity effects. *Journal of the Acoustical Society of America*, 1971, *49*, 1131–1139.

Arthur, R. M., Pfeiffer, R. R., & Suga, N. Properties of "two tone inhibtion" in primary auditory neurons. *Journal of Physiology* (London), 1971, *212*, 593–609.

Benevento, L. A., & Coleman, P. D. Responses of single cells in cat inferior colliculus to binaural click stimuli: Combinations of intensity levels, time differences, and intensity differences. *Brain Research*, 1970, *17*, 387–405.

Bobbin, R. P., & Thompson, M. H. Effects of putative transmitters on afferent cochlear transmission. *Annals of Otology, Rhinology, and Laryngology*, 1978, *87*, 185–191.

Boer, E. de. Reverse correlation. II Initiation of the nerve impulses in the inner ear. *Proceedings of the Koninklijke Nederlandse Akademie van Wetenschappen*, 1969, *72*, 129–151.

Borg, E. Efferent inhibition of afferent acoustic activity in the unanesthetized rabbit. *Experimental Neurology*, 1971, *31*, 301–312.

Boudreau, J. C., & Tsuchitani, C. Cat superior olive S-segment cell discharge to tonal stimulation. In W. D. Neff (Ed.), *Contributions to sensory physiology* (Vol. 4). New York: Academic Press, 1970.

Britt, R. H. Intracellular study of synaptic events related to phase-locking responses of cat cochlear nucleus cells to low frequency tones. *Brain Research*, 1976, *112*, 313–327.

Britt, R., & Starr, A. Synaptic events and discharge patterns of cochlear nucleus cells. I. Steady-frequency tone bursts. *Journal of Neurophysiology*, 1976, *39*, 162–178.

Brownell, W. E. Organization of the cat trapezoid body and the discharge characteristics of its fibers. *Brain Research,*, 1975, *94*, 413–433.

Brugge, J., Anderson, D., Hind, J., & Rose, J. Time structure of discharges in single auditory nerve fibers of the squirrel monkey in response to complex periodic sounds. *Journal of Neurophysiology*, 1969, *32*, 386–401.a

Brugge, J., Dubrovsky, N. A., Aitkin, L. M., & Anderson, D. J. Sensitivity of single neurons in the auditory cortex of cat to binaural stimulation: Effects of varying interaural time and intensity. *Journal of Neurophysiology*, 1969, *32*, 1005–1024.b

Brugge, J., Anderson, D. J., & Aitkin, L. J. Responses of neurons in the dorsal nucleus of the lateral lemniscus of the cat to binaural tonal stimulation. *Journal of Neurophysiology*, 1970, *33*, 441–458.

Caspary, D. Classification of subpopulations of neurons in the cochlear nuclei of the kangaroo rat. *Experimental Neurology*, 1972, *37*, 131–151.

Chow, K. L. Numerical estimates of auditory central nervous system of the rhesus monkey. *Journal of Comparative Neurology*, 1951, *95*, 159–175.

Clopton, B. M., Winfield, J. A., & Flammino, F. J. Tonotopic organization: Review and analysis. *Brain Research*, 1974, *76*, 1–20.

Colwell, S. A., & Merzenich, M. M. Corticothalamic projections from physiologically defined loci in AI in the cat. *Journal of Comparative Neurology*, (in press).

Comis S. D., & Leng, G. Action of putative neurotransmitters in the guinea pig cochlea. *Experimental Brain Research,* 1979, *36,* 119–128.

Dallos, P. *The auditory periphery.* New York: Academic Press, 1973.

Dallos, P. Comment on Rhode, W.S.: Some observations on two tone interaction measured with the Mössbauer effect. In E. F. Evans & J. P. Wilson (Eds.), *Psychophysics and physiology of hearing.* London: Academic Press, 1977.

Dallos, P., & Cheatham, M. A. Compound action potential (AP) tuning curves. *Journal of the Acoustical Society of America,* 1976, *59,* 591–597.

Desmedt, J. E. Auditory-evoked potentials from cochlea to cortex as influenced by activation of the efferent olivo-cochlear bundle. *Journal of the Acoustical Society of America,* 1962, *34,* 1478–1496.

Diamond, I. T., Jones, E. G., & Powe, T. P. S. The projection of the auditory cortex upon diencephalon and brain stem of the cat. *Brain Research,* 1969, *15,* 305–340.

Eggermont, J. J. Compound action potential tuning curves in normal and pathological ears. *Journal of the Acoustical Society of America,* 1977, *62,* 1247–1251.

Erulkar, S. D. The responses of single units of the inferior colliculus of the cat to acoustic stimulation. *Proceedings of the Royal Society of London* (Biological Sciences), 1959, *150,* 336–355.

Erulkar, S. D. Physiological studies of the inferior colliculus and medial geniculate complex. In W. D. Keidel & W. D. Neff (Eds.), *Handbook of sensory physiology* (Vol. V/2). New York: Springer-Verlag, 1975.

Erulkar, S. D., Nelson P. G., & Bryan, J. S. Experimental and theoretical approaches to neural processing in the central auditory pathway. *Contributions to Sensory Physiology* (Vol. 3). New York: Academic Press, 1968.

Evans, E. F. The frequency response and other properties of single fibers in the guinea pig cochlear nerve. *Journal of Physiology* (London), 1972, *226,* 263–287.

Evans, E. F. Normal and abnormal functioning of the cochlear nerve. *Symposium of the Zoological Society of London,* 1975, *37,* 133–165.a

Evans, E. F. Cochlear nerve and cochlear nucleus. In, W. D. Keidel & W. D. Neff (Eds.), *Handbook of Sensory Physiology* (Vol. 2). New York: Springer-Verlag, 1975. b

Evans, E. F., & Klinke R., Reversible effects of cyanide and furosemide on the tuning of single cochlear fibers. *Journal of Physiology* (London), 1974, *242,* 129–131.

Evans, E. F., & Nelson, P. G. On the relationship between the dorsal and ventral cochlear nucleus. *Experimental Brain Research,* 1973, *17,* 428–442.

Fernandez, C., & Karapas, F. The course and termination of the stria of Monakow and Held in cat. *Journal of Comparative Neurology,* 1967, *131,* 371–386.

Fex, J. Auditory activity in centrifugal and centripetal fibers in the cat. A study of a feedback system. *Acta Physiologica Scandinavica,* 1962 (Suppl.), *55,* 189.

Fex, J. Auditory activity in uncrossed centrifugal cochlear fibers in a cat. A study of a feedback system. *Acta Physiologica Scandinavica,* 1965, *64,* 43–57.

Fex J. Efferent inhibition in the cochlea related to hair cell dc activity: Study of post synaptic activity of the crossed olivo cochlear fibres in the cat. *Journal of the Acoustical Society of America,* 1967, *41,* 666–675.

Flock, A. Contractible proteins in haircells. *Hearing Research,* 1980, *2,* 411–412.

Flock A., & Cheung, H. C. Actin filaments in sensory hairs of inner ear receptor cells. *Journal of Cell Biology,* 1977, *75,* 339–343.

Furst, M., & Goldstein, J. L. Differences of CT ($2f_1 - f_2$) phase in psychophysical and physiological experiments. *Hearing Research,* 1980, *2,* 379–386.

Gacek, R. H., & Rasmussen, G. L. Fiber analysis in the statoacoustic nerve of guinea pig, cat and monkey. *Anatomical Record*, 1961, *139*, 455–463.

Galambos, R., & Davis H. Responses of single auditory nerve fibers to acoustic stimulation. *Journal of Neurophysiology*, 1943, *6*, 39–57.

Geisler, C. D., Rhode, W. S., & Hazelton, D. W. Responses of inferior colliculus neurons in the cat to binaural acoustic stimuli having wideband spectra. *Journal of Neurophysiology*, 1969, *32*, 960–974.

Geisler, C. D., Rhode, W. S., & Kennedy, D. T. Responses to tonal stimuli of single auditory nerve fibers and their relationship to basilar membrane motion in the squirrel monkey. *Journal of Neurophysiology*, 1974, *37*, 1156–1172.

Gerstein, G. L., Butler, R. A., & Erulkar, S. D. Excitation and inhibition in cochlear nucleus. I. Tone burst stimulation. *Journal of Neurophysiology*, 1968, *31*, 526–536.

Glattke, T. J. Unit responses of the cat cochlear nucleus to amplitude modulated stimuli. *Journal of the Acoustical Society of America*, 1969, *45*, 419–425.

Goblick, T. J., & Pfeiffer. R. R. Time domain measurements in cochlear nonlinearities using combination click stimuli. *Journal of the Acoustical Society of America*, 1969, 924–938.

Godfrey, D. A., Kiang, N.-Y. S., & Norris, B. E. Single unit activity in the posteroventral cochlear nucleus of the cat. *Journal of Comparative Neurology*, 1975, *162*, 247–268. a

Godfrey, D. A., Kiang, N.-Y. S., & Norris, B. E. Single unit activity in the dorsal cochlear nucleus of the cat. *Journal of Comparative Neurology*, 1975, *162*, 269–284. b

Goldberg, J. M. Physiological studies of auditory nuclei of the pons. In W. D. Keidel & W. D. Neff (Eds.), *Handbook of sensory physiology* (Vol. V/2), Berlin: Springer-Verlag, 1975.

Goldberg, J. M., & Brown, P. B. Responses of binaural neurons of dog superior olivary complex to dichotic tonal stimuli: Some physiological mechanisms of sound localization. *Journal of Neurophysiology*, 1969, *32*, 613–636.

Goldberg, J. M., & Brownell, W. E. Discharge characteristics of neurons in anteroventral and dorsal nuclei of the cat. *Brain Research*, 1973, *64*, 34–54.

Goldberg, J. M., Adrian, H. O., & Smith, F. D. Response of neurons in the superior olivary complex of the cat to acoustic stimuli of long duration. *Journal of Neurophysiology*, 1964, *27*, 706–749.

Goldstein, J. L. Auditory nonlinearity. *Journal of the Acoustical Society of America*, 1967, *41*, 676–689.

Goldstein, J. L. Aural combination tones. In *Frequency Analysis and Periodicity Detection in Hearing*. Pp. 230–245. Leiden: A. W. Sijthoff, 1970.

Goldstein, J. L., & Kiang, N.-Y. S. Neural correlates of the aural combination tones $2f_1 - f_2$. *Proceedings of IEEE*, 1968, *56*, 981–992.

Greenwood, D. D., & Maruyama, N. Excitatory and inhibitory response areas of auditory neurons in the cochlear nucleus. *Journal of Neurophysiology*, 1965, *28*, 863–892.

Greenwood, D. P., Merzenich, M. M., & Roth, G. Some preliminary observations on inter-relations between two-tone suppression and combination tone driving in anteroventral cochlear nucleus of the cat. *Journal of the Acoustical Society of America*, 1976, *59*, 607–633.

Guinan, J. J., & Norris, B. E., Single auditory units in the superior olivary complex II. Locations of unit categories and tonotopic organization. *International Journal of Neuroscience*, 1972, *4*, 147–166. a

Guinan, J. J., Guinan, S. S., & Norris, B. E. Single auditory units in the superior olive complex I. Responses to sounds and classifications based on physiological properties. *International Journal of Neuroscience*, 1972, *4*, 101–120. b

Guth, P. S., Norris, C. H., & Bobbin, R. P. The pharmacology of transmission in the peripheral auditory system. *Pharmacological Reviews*, 1976, *28*, 95–125.

Hall, J. L. Binaural interaction in the accessory superior-olivary nucleus of the cat. *Journal of the Acoustical Society of America*, 1965, *37*, 814–823.

Harris, D. M. Action potential suppression, tuning curves, and thresholds: Comparison with single fiber data. *Hearing Research*, 1979, 1, 133–154.

Harrison, J. M., & Feldman, M. L. Anatomical aspects of the cochlear nucleus and superior olivary complex. In W. D. Neff (Ed.), *Contributions to sensory physiology* (Vol. 4). New York: Academic Press, 1970.

Harrison, J. M., & Howe, M. E. Anatomy of the descending auditory system (mammalian). In W. D. Keidel & W. D. Neff (Eds.), *Handbook of sensory physiology* (Vol. V/1). Berlin: Springer-Verlag, 1974.

Harrison, J. M., & Irving, R. The organization of the posterior ventral cochlear nucleus. *Journal of Comparative Neurology*, 1966, *126*, 391–403.

Harrison, J. M., & Warr, W. B. A study of the cochlear nuclei and ascending pathways of the medulla. *Journal of Comparative Neurology*, 1962, *119*, 341–380.

Helle, R. Amplitude und Phase des im Gehor gebildeten Differenztones dritter Ordmung. *Acustica*, 1969, *22*, 74–87.

Hind, J. E., Goldberg, J. M., Greenwood, D. D., & Rose, J. E. Some discharge characteristics of single neurons in the inferior colliculus of the cat. II. Timing of the discharges and observations on binaural stimulation. *Journal of Neurophysiology*, 1963, *26*, 321–341.

Hind, J. E., Anderson, D. J., Brugge, J. F., & Rose, J. E. Coding of information pertaining to paired low-frequency tones in single auditory nerve fibers of the squirrel monkey. *Journal of Neurophysiology*, 1967, *30*, 794–816.

Hind, J. E., Rose, J. E., Brugge, J. F., & Anderson, D. J. Two-tone masking effects in squirrel monkey auditory nerve fibers. In R. Plomp & G. F. Smoorenburg (Eds.), *Frequency analysis and periodicity detection in hearing*. Leiden: A. W. Sijthoff, 1970.

Imig, T. J., & Adrian, H. O. Binaural columns in the primary field (AI) of cat auditory cortex. *Brain Research*, 1978, *138*, 241–257.

Imig, T. J., & Brugge, J. F. Sources and terminations of callosal axons related to binaural and frequency maps in primary auditory cortex of the cat. *Journal of Comparative Neurology*, 1978, *182*, 637–660.

Jungert, S. Auditory pathways in the brainstem. A neurophysiological study. *Acta Otolaryngologica* (Stockholm), 1958 (Suppl. *138*).

Kallert, S. Telemetrische Mikroelektrodenuntersuchungen am Corpus geniculatum mediale der wachen Katze. Habil.-Schr Erlangen, 1974.

Kallert, S., & Keidel, W. D. Telemetrical microelectrode study of the upper parts of the auditory pathway in free moving cats. *Pfluegers Archiv. European Journal of Physiology*, 1973, *343*, 79.

Kallert, S., David, E., Finkenzeller, P., & Keidel, W. D. Two different neuronal discharge periodicities in the acoustic channel. In R. Plomp & G. F. Smoorenburg (Eds.), *Frequency analysis and periodicity detection in hearing*. Leiden: A. W. Sijthoff, 1970.

Kane, E. S. Primary afferents and the cochlear nucleus. In R. F. Naunton & C. Fernandez (Eds.), *Evoked electrical activity in the auditory system*. New York: Academic Press, 1978.

Katsuki, Y. Integrative organization of the thalamic and cortical auditory centers. In D. P. Purpura (Ed.), *The thalamus*. New York: Columbia University Press, 1966.

Katsuki, Y., Sumi, T., Uchiyama, H., & Watanabe, T. Electric responses of auditory neurons in cat to sound stimulation. *Journal of Neurophysiology*, 1958, *21*, 569–588.

Katsuki, Y., Suga, N., & Karmo, Y. Neural mechanisms of the peripheral and central auditory system in monkey. *Journal of the Acoustical Society of America*, 1962, *34*, 1396–1410.

Kemp, D. T. Evidence of mechanical nonlinearity and frequency selective wave amplification in the cochlea. *Archives of Oto-Rhino-Laryngology*, 1979, *224*, 37–45.

Kiang, N.-Y. S. Stimulus representation in the discharge patterns of auditory neurons. In E. L. Eagles (Ed.), *The nervous system* (Vol. 3), *Human communication and its disorders*. New York: Raven Press, 1975.

Kiang, N.-Y. S., Watanabe, T., Thomas, E. C., & Clark, L. F. *Discharge patterns of single fibers in the cat's auditory nerve*. Research Monograph 35, Cambridge, Massachusetts: MIT Press, 1965. a

Kiang, N.-Y. S., Pfeiffer, R. R., Warr, W. B., & Backus, A. S. N. Stimulus coding in the cochlear nucleus. *Annals of Otology, Rhinology, and Laryngology*, 1965, *74*, 463–485. b

Kiang, N.-Y. S., Sachs, M. B., & Peake, W. T. Shapes of tuning curves for single auditory nerve fibers. *Journal of the Acoustical Society of America*, 1967, *42*, 1341–1342.

Kiang, N.-Y. S., Morest, D. K., Godfrey, D. A., Guinan, J. J., & Kane, E. C. Stimulus coding at caudal levels of the cat's auditory nervous system. 1. Response characteristics of single units. In A. R. Møller (Ed.), *Basic mechanisms in hearing*. New York: Academic Press, 1973.

Kiang, N.-Y. S., & Moxon, E. C. Tails of tuning curves of auditory nerve fibers. *Journal of the Acoustical Society of America*, 1974, *55*, 620–630.

Kim, D. O. Cochlear mechanics: Implications of electrophysiological and acoustical observations. *Hearing Research*, 1980, *2*, 297–317.

Kim, D. O., Siegel, J. H., & Molnar, C. D. Cochlear nonlinear phenomena in two-tone responses. *Scandinavian Audiology*, 1979 (Suppl.), *9*, 63–81.

Kim, D. O., Molnar, C. E., & Matthews, J. W., Cochlear mechanics: non-linear behavior in two-tone responses as reflected in cochlear-nerve-fiber responses and in ear-canal sound pressure. *Journal of the Acoustical Society of America*, 1980, *67*, 1704–1721.

Klinke, R., & Galley, N. Efferent innervation of vestibular and auditory receptors. *Physiological Reviews*, 1974, *54*, 316–357.

Koningsmark, B. W. Neuroanatomy of the auditory system. *Archives of Otolaryngology*, 1973, *98*, 397–413.

Larsell, O. *Anatomy of the nervous system*. New York: D. Appleton- Century Co., 1951.

Lavine, R. A. Phase locking in response in single neurons in cat cochlear nucleus to low-frequency tonal stimuli. *Journal of Neurophysiology*, 1971, *34*, 467–483.

Lazorthes, G., Lacomme, Y., Gaubert, J., & Planel, H. La constitution du nerf auditif. *La Press Medicale*, 1961, *69*, 1067–1068.

Legoux, J. P., Remond, M. C., & Greenbaum, H. B., Interference and two tone inhibition. *Journal of the Acoustical Society of America*, 1973, *53*, 409–419.

Liberman, M. D., & Kiang, N.-Y. S. Acoustic trauma in cats. *Acta Otolaryngologica* (Stockholm), 1978, (Suppl. *358*), 63.

Lórente de No, R. Anatomy of the eighth nerve. I. The central projections of the nerve endings of the internal ear. *Laryngoscope.* 1933, *43*, 1–38.a

Lórente de No, R. Anatomy of the eighth nerve. III. General plan of structure of the primary cochlear nuclei. *Laryngoscope,* 1933, *43*, 327–350. b

Marmarelis, P. Z., & Marmarelis, V. Z. *Analysis of physiological systems.* New York: Plenum Press, 1978.

Mast, T. E. Binaural interaction and contralateral inhibition in dorsal cochlear nucleus of chinchilla. *Journal of Neurophysiology,* 1970, *33,* 108–115.

Merzenich, M. M., & Brugge, J. F. Representation of the cochlear partition on the superior temporal plane of the macaque monkey. *Brain Research,* 1973, *50,* 275–296.

Merzenich, M. M., & Kaas, J. Principles of organization of sensory- perceptual systems in mammals. In J. M. Sprague & A. M. Epstein (Eds.), *Progress in psychobiology and physiological psychology* (Vol. IX). New York: Academic Press, 1980.

Merzenich, M. M., Knight, P. L., & Roth, G. L. Representation of cochlea within primary auditory cortex in the cat. *Journal of Neurophysiology,* 1975, *38,* 231–249.

Merzenich, M. M., Roth, R. A., Knight, P. L., & Colwell, S. A. Some basic features of organization of the central auditory system. In E. F. Evans & J. P. Wilson (Eds.), *Psychophysics and physiology of hearing.* London: Academic Press, 1977.

Merzenich, M. M., Anderson, R. A., & Middlebrooks, J. H. Functional and topographic organization of the auditory cortex. *Experimental Brain Research.* Suppl. 2, 1979, 61–75.

Middlebrooks, J. C., Dykes, R. W., & Merzenich, M. M. Binaural response-specific bands in primary auditory cortex (AI) of the cat: Topographical organization orthogonal to isofrequency contours. *Brain Research,* 1980, *181,* 31–48.

Møller, A. R. Unit responses in cochlear nucleus of the rat to pure tones. *Acta Physiologica Scandinavica,* 1969, *75,* 530–541. a

Møller, A. R. Unit responses in the rat cochlear nucleus to repetitive transient sounds. *Acta Physiologica Scandinavica,* 1969, *75,* 542–551. b

Møller, A. R. Unit responses in the cochlear nucleus of the rat to sweep tones. *Acta Physiologica Scandinavica,* 1969, *76,* 503–512.c

Møller, A. R. Unit responses in cochlear nucleus of the rat to noise and tones. *Acta Physiologica Scandinavica,* 1970, *78,* 289–298.

Møller, A. R. Unit responses in the rat cochlear nucleus to tones of rapidly varying frequency and amplitude. *Acta Physiologica Scandinavica,* 1971, *81,* 540–556.

Møller, A. R. Coding of sounds in lower levels of the auditory system. *Quarterly Review of Biophysics,* 1972, *5,* 59–155. a

Møller, A. R. Coding of amplitude and frequency modulated sounds in the cochlear nucleus of the rat. *Acta Physiologica Scandinavica,* 1972, *86,* 223–238.b

Møller, A. R. Responses of units in the cochlear nucleus to sinusoidally amplitude-modulated tones. *Experimental Neurology,* 1974, *45,* 104–117.a

Møller, A. R. Coding of amplitude and frequency modulated sounds in the cochlear nucleus. *Acustica,* 1974, *31,* 292–299.b

Møller, A. R. Dynamic properties of excitation and inhibition in the cochlear nucleus. *Acta Physiologica Scandinavica,* 1975, *93,* 442–454.a

Møller, A. R. Latency of unit responses in cochlear nucleus determined in two different ways. *Journal of Neurophysiology,* 1975, *38,* 818–821.b

Møller, A. R. Dynamic properties of excitation and two-tone inhibition of the cochlear nucleus studied under amplitude-modulated tones. *Experimental Brain Research,* 1976, *25,* 307–321.a

Møller, A. R. Dynamic properties of the responses of single neurons in the cochlear nucleus. *Journal of Physiology* (London), 1976, *259*, 63–83. b

Møller, A. R. Frequency selectivity of single auditory nerve fibers in response to noise stimuli. *Journal of the Acoustical Society of America*, 1977, *62*, 135–142.

Moore, R. Y., & Goldberg, J. M. Ascending projections of the inferior colliculus in the cat. *Journal of Comparative Neurology*, 1963, *121*, 109.

Morest, D. K. The laminar structure of the inferior colliculus of the cat. *Anatomical Record*, 1964, *148*, 390–391. a

Morest, D. K. The neuronal architecture of the medial geniculate body of the cat. *Journal of Anatomy*, 1964, *98*, 611–630. b

Morest, D. K. The cortical structure of the inferior quadrigeminal lamina of the cat. *Anatomical Record*, 1966, *154*, 389–390. a

Morest, D. K. The noncortical neuronal architecture of the inferior colliculus of the cat. *Anatomical Record*, 1966, *154*, 477. b

Morest, D. K. The collateral system of the media nucleus of the trapezoid body of the cat, its neuronal architecture and relation to the olivo-cochlear bundle. *Brain Research*, 1968, *9*, 288–311.

Morest, D. K., Kiang, N.-Y. S., Kane, E. C., Guinan, J. J., & Godfrey, D. A. Stimulus coding of caudal levels of the cat's auditory nervous system. II. Patterns of synoptic organization. In A. R. Møller (Ed.), *Basic mechanisms in hearing*, New York: Academic Press, 1973.

Mountain, D. C. Changes in endolymphatic potential and crossed olivo cochlear bundle stimulation alter cochlear mechanics. *Science*, 1980, *210*,71–72.

Mountcastle, V. B. *Medical physiology* (Vol. 1). St. Louis: C. V. Mosby Co., 1974.

Moushegian, G., Rupert, A. L., & Whitcomb, M. A. Brain stem neuronal response patterns to nonaural tones. *Journal of Neurophysiology*, 1964, *27*, 1174–1191.

Newman, J. D. & Symmes, D. Feature detection by single units in squirrel monkey auditory cortex. In O. Creutzfeldt, H. Scheich, & C. Schreiner (Eds.), *Hearing mechanisms and speech. Experimental Brain Research*, 1979 (Suppl.) 2.

Nomoto, M., Suga, N., & Katsuki, Y. Discharge pattern and inhibition of primary nerve fibers in the monkey. *Journal of Neurophysiology*, 1964, *27*, 768–787.

Osen, K. K. Cytoarchitecture of the cochlear nuclei in the cat. *Journal of Comparative Neurology*, 1969, *136*, 453–483.

Osen, K. K. Course and termination of the primary afferents in cochlear nuclei of the cat. *Archivio Italiano di Otologia, Rinologia e Laringologia*, 1970, *108*, 21–51.

Pfeiffer, R. R. *Electrophysiological response characteristics of single units in the cochlear nucleus of the cat*. Unpublished doctoral dissertation, Massachusetts Institute of Technology, 1963.

Pfeiffer, R. R. Classification of response patterns of spike discharges for units in the cochlear nucleus: Tone-burst stimulation. *Experimental Brain Research*, 1966, *I*, 220–235.

Pierson, M., & Møller, A. R. Some dualistic properties of the cochlear microphonics. *Hearing Research*, 1980, *2*, 135–149. a

Pierson, M., & Møller, A. R. Effect of modulation of basilar membrane position on the cochlear microphonic. *Hearing Research*, 1980, *2*, 151–162. b

Rhode, W. S. Some observations on two-tone interaction measured with the Mössbauer effect. In E. F. Evans & J. P. Wilson (Eds.), *Psychophysics and physiology of hearing*. New York: Academic Press, 1977.

Rockel, A. J., & Jones, E. G. The neuronal organization of the inferior colliculus of the adult cat I. Central nucleus. *Journal of Comparative Neurology*, 1973, *147*, 11-60. a

Rockel, A. J., & Jones, E. G. Observations of the fine structure of the central nucleus of the inferior colliculus of the cat. *Journal of Comparative Neurology*, 1973, *147*, 61-92. b

Rockel, A. J., & Jones, E. G. The neural organization of the inferior colliculus of the adult cat II. The pericentral nucleus. *Journal of Comparative Neurology*, 1973, *149*, 301-334. c

Romand, R. Intracellular recordings of "chopper responses" in the cochlear nucleus of the cat. *Hearing Research*, 1979, *1*, 95-99.

Rose, J. E. The cellular structure of the auditory region of the cat. *Journal of Comparative Neurology*, 1949, *91*, 409-440.

Rose, J. E., Galambos, R., & Hughes, J. R. Microelectrode studies of the cochlear nuclei in the cat. *Bulletin of the Johns Hopkins Hospital*, 1959, *104*, 211-251.

Rose, J. E., Greenwood, D. D., Goldberg, J. M., & Hind, J. E. Some discharge characteristics of single neurons in the inferior colliculus of cat. I.: Tonotopic organization, relation of spike counts to tone intensity and firing patterns of single elements. *Journal of Neurophysiology*, 1963, *26*, 294-320.

Rose, J. E., Gross, N. B., Geisler, C. D., & Hind, J. E. Some neural mechanisms in the inferior colliculus of the cat which may be relevant to localization of a sound source. *Journal of Neurophysiology*, 1966, *29*, 288-314.

Rose, J. E., Brugge, J. F., Anderson, D. J., & Hind, J. E. Phase-locked responses to low-frequency tones in single auditory nerve fibers of the squirrel monkey. *Journal of Neurophysiology*, 1967, *30*, 769-793.

Rose, J. E., Hind, J. E., Anderson, D. J., & Brugge, J. F. Some effects of stimulus intensity on response of auditory nerve fibers in the squirrel monkey. *Journal of Neurophysiology*, 1971, *34*, 685-699.

Rouiller, E., de Ribaupierre, Y., & de Ribaupierre, F. Phase-locked responses to low frequency tones in the medial geniculate body. *Hearing Research*, 1979, *1*, 213-226.

Rupert, A. L., & Moushegian, G. Neural responses of kangaroo rat ventral cochlear nucleus to low frequency tones. *Experimental Neurology*, 1970, *26*, 84-102.

Rupert A., Moushegian, G., & Galambos, R. Units responses to sound from auditory nerve of the cat. *Journal of Neurophysiology*, 1963, *26*, 449-465.

Russell, I. J., & Sellick, P. M. Intracellular studies of hair cells in the mammalian cochlea. *Journal of Physiology* (London), 1978, *284*, 261-290.

Sachs, M. B. Stimulus-response relation for auditory-nerve fibers: Two-tone stimuli. *Journal of the Acoustical Society of America*, 1969, *45*, 1025-1036.

Sachs, M. B., & Abbas, P. J. Rate versus level functions for auditory-nerve fibers in cats: Tone burst stimuli. *Journal of the Acoustical Society of America*, 1974, *56*, 1835-1847.

Sachs, M. B., & Kiang, N.-Y. S. Two-tone inhibition in auditory nerve fibers. *Journal of the Acoustical Society of America*, 1968, *43*, 1120-1128.

Sachs, M. B., Young, E. D., & Lewis, R. H. Discharge patterns of single fibers in the pigeon auditory nerve. *Brain Research*, 1974, *70*, 431-447.

Sando, I. The interrelationships of the cochlear nerve fibers. *Acta Otolaryngologica* (Stockholm), 1965, *59*, 417-436.

Sellick, P. M., & Russell, I. J. Two-tone suppression in cochlear hair cells. *Hearing Research*, 1979, *1*, 227-236.

Sellick, P. M., & Russell, I. J. The responses of inner hair cells to basilar membrane velocity during low frequency auditory stimulation in the guinea pig cochlea. *Hearing Research*, 1980, *2*, 439-445.

Siegel, J. H. *Effects of altering the organ of Corti on cochlear distortion products* ($f_2 - f_1$) and $2f_1 - f_2$. Unpublished doctoral dissertation, Washington University, 1978.

Siegel, J. H., Kim, D. O., & Molnar, C. E. Effects of altering organ of Corti on cochlear distortion products $f_2 - f_1$ and $2f_1 - f_2$. *Journal of Neurophysiology*, 1982, *47*, 303–328.

Sinex, D. G., & Geisler, C. D. Auditory fiber responses to frequency-modulated tones. *Hearing Research*, 1981, *4*, 127–148.

Smoorenburg, G. F. Combination tones and their origin. *Journal of the Acoustical Society of America*, 1972, *52*, 615–632.

Smoorenburg, G. F. Effect of temporary threshold shift on combination tone generation and a two tone suppression. *Hearing Research*, 1980, *2*, 357–368.

Smoorenburg, F. G., Gibson, M. M., Kitzes, L. M., & Rose, J. E. Correlates of combination tones observed in response of neurons in the anteroventral cochlear nucleus of the cat. *Journal of the Acoustical Society of America*, 1976, *59*, 945–962.

Snider, R. S., & Stowell, A. Receiving areas of tactile, auditory and visual systems in the cerebellum. *Journal of Neurophysiology*, 1944, *7*, 331–357.

Spoendlin, H. The organization of the cochlear receptor. *Advances in Otorhinolaryngology*, 1966, *13*.

Spoendlin, H. Structural basis of peripheral frequency analysis. In R. Plomp & G. F. Smoorenburg (Eds.) *Frequency analysis and periodicity detection in hearing*. Leiden: A. W. Sijthoff, 1970.

Star, A., & Britt, R. Intracellular recordings from cat cochlear nucleus during tone stimulation. *Journal of Neurophysiology*, 1970, *33*, 137–147.

Stotler, W. A. An experimental study of the cells and connections of the superior olivary complex of the cat. *Journal of Comparative Neurology*, 1953, *98*, 401–423.

Suga, N. *Cortical representation of biosonar information in the mustached bat*. The 28th International Congress of Physiological Science, Publishing House of Hungarian Academy of Science, 1981.

Suga, N., Kuzirai, K., & O'Neill, W. E. How biosonar information is represented in the bat cerebral cortex. In J. Syka & L. Aitkins (Eds.), *Neuronal mechanisms in hearing*. New York: Plenum, 1981.

Tamar, H. *Principles of sensory physiology*. Springfield, Illinois: Charles C Thomas, 1972.

Tasaki, I. Nerve impulses in individual nerve fibers of the guinea pig. *Journal of Neurophysiology*, 1954, *17*, 97–122.

Tsuchitani, C. Functional organization of lateral cell groups of cat superior olivary complex. *Journal of Neurophysiology*, 1977, *40*, 296–318.

Tsuchitani, C., & Boudreau, J. C. Single unit analysis of cat superior olive S-segment with tonal stimuli. *Journal of Neurophysiology*, 1966, *29*, 684–697.

Tsuchitani, C., & Boudreau, J. C. Encoding of stimulus frequency and intensity by cat superior olive S-segment cells. *Journal of the Acoustical Society of America*, 1967, *42*, 794–805.

Tunturi, A. R. Physiological determination of the arrangement of afferent connections to the middle ectosylvian area in the dog. *American Journal of Physiology*, 1950, *162*, 489–502.

van Noort, J. *The structure and connections of the inferior colliculus*. Assen: Van Gorcum, Ltd., 1969.

Warr, W. B. Fiber degeneration following lesions in the anterior ventral cochlear nucleus of the cat. *Experimental Neurology*, 1966, *14*, 453–474.

Warr, W. B. Fiber degeneration following lesion in the posterior ventral cochlear nucleus of the cat. *Experimental Neurology* 1969, *23*, 140–155.

Warr, W. B. The olivocochlear bundle: Its origin and terminations in the cat. In R. F. Naunton & C. Fernandez (Eds.), *Evoked electrical activity in the auditory nervous system*. New York: Academic Press, 1978.

Watanabe, T. Funneling mechanisms in hearing. *Hearing Research*, 1979, *1*, 111–119.

Wever, E. G., & Lawrence, M. *Physiological acoustics*. Princeton, New Jersey: Princeton University Press, 1954.

Whitfield, I. C. *The auditory pathway*. London: Arnold, 1967.

Wiederhold, M. L. Variations in the effects of electrical stimulation of the crossed olivocochlear bundle on cat single auditory nerve fiber response to tone bursts. *Journal of the Acoustical Society of America*, 1970, *48*, 966–977.

Wiederhold, M. L., & Kiang, N.-Y. S. Effects of electrical stimulation of the crossed olivo cochlear bundle on single auditory nerve fibers in the cat. *Journal of the Acoustical Society of America*, 1970, *48*, 950–965.

Wiederhold, M. L., & Peake, W. T. Efferent inhibition of auditory-nerve responses: Dependence on acoustic-stimulus parameters. *Journal of the Acoustical Society of America*, 1966, *40*, 1427–1430.

Winer, J. A., Diamond, I. T., & Raczkowsky, D. Subdivisions of auditory cortex of the cat, the retrograde transport of horseradish peroxidase to medial geniculate body and posterior thalamic nuclei. *Journal of Comparative Neurology*, 1977, *176*, 387–418.

Woolsey, C. N. Organization of cortical auditory system: A review and a synthesis. In G. Rasmussen & W. Windle (Eds.), *Neural mechanisms of the auditory and vestibular systems*. Springfield, Illinois: Charles C Thomas, 1960.

Woolsey, C. N., & Walzl, E. M. Topical projection of nerve fibers from local regions of the cochlea in the cerebral cortex of the cat. *Bulletin of the Johns Hopkins Hospital*, 1942, *71*, 315–344.

Zwicker, E. Nonmonotonic behavior of $(2f_1 - f_2)$ explained by a saturation-feedback model. *Hearing Research*, 1980, *2*, 513–518.

Zwislocki, J. J., & Kletsky, E. J. Outer hair cells: Sharpness of tuning. *Acta Otolaryngologica*, 1981, *91*, 481–485.

CHAPTER **3**

Frequency Analysis in
the Peripheral Auditory System

Introduction

In the previous two chapters, the general anatomy and function (physiology) of the ear and the auditory nervous system were described. In this chapter, we will consider in some detail the neurophysiological basis for discrimination of frequency and time.

Frequency discrimination and frequency analysis are two concepts that have become fundamental in the study of psychoacoustics. Frequency discrimination refers to the ability to distinguish two sounds from each other on the basis of their frequencies when the two sounds are not presented simultaneously. Frequency analysis refers to the ability to discriminate one spectral component in a complex sound that contains several spectral components. Frequency analysis is usually thought of as the ability to resolve the different components that comprise a sound.

Studies of frequency discrimination have usually used pure tones, and the results indicate the number of hertz difference needed for an observer to hear that the two tones are different. Both frequency discrimination and frequency analysis are commonly associated with properties of the ear and with properties of the peripheral auditory nervous system.

The mechanics responsible for frequency discrimination are not clear. Such discrimination may be based on spectral analysis in the inner ear, or it may be related to the time structure of sound and thus depend on temporal analysis performed by the auditory nervous system.

Hypotheses have adhered to one of two general principles, in which discrimination is based on either spectral or temporal analysis. The spectral hypothesis, also known as the *place principle,* assumes that tones with different frequencies give rise to maximal vibration amplitudes at different places on the basilar membrane. More specifically, the locations of the maxima are directly related to the frequency of a pure tone. A high-frequency tone is assumed to give rise to maximum amplitude of vibration of the basilar membrane near the base of the cochlea; a low-frequency tone will produce maximal vibration amplitude near the apex of the cochlea. For a complex sound, there is assumed to be a separation of spectral components along the basilar membrane. The hypothesis of spectral analysis in the ear, originally formulated by von Helmholtz (1863), has strongly influenced the thinking and experimental design of psychoacoustics.

The place hypothesis of hearing may be regarded as an auditory version of Müller's theory on "the specific energies of the nerves" (Müller, 1837). This doctrine was originally applied to different sensory nerves, and it states that it is not the *type of stimulation* that determines whether we experience an auditory, visual, or olfactory sensation, but it is the *type of nerve* that is stimulated that determines the type of sensation. According to this doctrine, the sensation that results from stimulation of the auditory nerve will always be auditory, and stimulation from the optic nerve will be visual.

Müller's hypothesis can also be extended to apply to stimulation of the various auditory nerve fibers. Whatever the stimulation of nerve fibers may be, the sensation associated with the stimulation of different fibers of the auditory nerve will be in accordance with the *location* of the fibers along the basilar membrane. Thus, stimulation of fibers ending in the apical part of the auditory nerve will result in a sensation of a low-frequency sound, stimulation of nerve fibers ending in the basal part of the cochlea will give rise to a sensation of a high-frequency sound, and so on.

Records from single nerve elements support the place principle of hearing. Thus, the discharge rate in single auditory nerve fibers has been found to be a function of the frequency of pure tones used as stimuli (see Kiang, Watanabe, Thomas, & Clark, 1965). That is, a pure tone will give rise to a larger discharge rate in a certain population of nerve fibers than it will in others.

Distribution of discharge rates among nerve fibers was therefore assumed to reflect the distribution of vibration amplitude along the basilar membrane. By locating the nerve fiber that has the highest discharge rate, the point with the highest amplitude can be identified. Studies of neurons at higher levels revealed a similar frequency selectivity. The place principle assumes that the central auditory nervous system recognizes that group of fibers that carries the highest discharge rate and thereby determines the frequency of the sound (Figure 3.1). The tonotopic organization of neurons in the various nuclei of the ascending auditory pathway, including the auditory cortex, supports the place principle of frequency analysis in the ear. Results of the foregoing studies lead investigators to assume that spectral analysis is the basis for frequency analysis in the ear in general. However, these results were all obtained using very simple sounds, for example, pure tones with constant frequencies and intensities presented one at a time without any other sounds. This is a situation that is very different from the normal situation, since natural sounds contain many frequency components at the same time, and both intensity and spectral composition vary more or less rapidly.

In the past, the majority of studies on spectral analysis in the auditory periphery have focused on the frequency selectivity of a single point along the basilar membrane or of single auditory nerve fibers or nerve cells. However, it is not known whether this ability to separate different frequencies plays any great role in discrimination of complex sounds, and it is not obvious that the perceptually important aspects of the analysis are performed in the auditory periphery. It has been claimed that the auditory periphery performs a *Fourier analysis* of signals. It should be pointed out that Fourier analysis is a mathematical operation that separates a complex

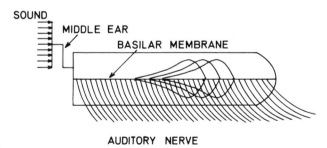

FIGURE 3.1. Schematic illustration of the frequency selectivity of the basilar membrane. The traveling wave envelope on the basilar membrane is indicated for three pure tones with different frequencies.

waveform into a series of sine waves, the amplitude and phase of which uniquely describes the complex waveform. Such an analysis cannot be made by any analog filter (bandpass filter), nor can it occur in the ear. It should be added that theories have been put forward suggesting that frequency selectivity may occur through spectral filtering *without* the resonance of the basilar membrane (see Huggins, 1952).

The *temporal principle,* on the other hand, is based on transmission of information about periodicity or time interval between individual waves of a periodic sound. It is assumed that this information about the time pattern of a sound is carried in the discharge pattern of individual nerve fibers. In its simplest form, the temporal principle assumes that individual nerve impulses are time locked to the waveform of a periodic sound. This is assumed to be relatively independent of localization of the nerve cells along the basilar membrane.

The temporal principle, as a basis for frequency discrimination, presumes that the central auditory nervous system is able to convert a time code into frequency, or to count the number of identical waves per unit of time. The temporal hypothesis for frequency discrimination also assumes that the form in which periodicity of primary auditory nerve fibers is coded is translated into another form somewhere in the nervous system. This translation is assumed to occur at a relatively peripheral level since synaptic transmission is likely to "blur" the time structure of the discharge pattern. So far, no such general transformation of the temporal pattern has been documented in neurophysiological studies.

The temporal principle of frequency discrimination is supported by the finding that periodicity—and even the wave shape to some extent—of a sound is contained in the discharge pattern of primary auditory nerve fibers (Figure 3.2). Coding of periodicity manifests itself as a clustering of discharges around certain phases (locations) of a period sound (e.g., a sine wave; see Chapter 2, p. 150). In this connection, it should be kept in mind that hair cells are sensitive to the deflection of the basilar membrane in only one direction. Therefore, the discharge pattern can only reflect the half-wave rectified version of the vibration of the basilar membrane. Coding of the periodicty of a sound in the auditory nerve has been shown to exist to a significant degree only for frequencies below 4–5 kHz. If this finding is correct, it naturally limits the temporal principle to frequencies below 5 kHz. Thus, the well-documented ability of the ear to discriminate frequencies above 5 kHz indicates that a different principle is at work.

Similar to what was pointed out regarding spectral selectivity (place principle), we do not know the perceptual importance of the coding of the

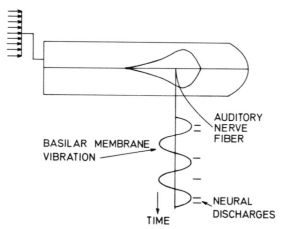

FIGURE 3.2. Schematic illustration of the phase-locking between the vibration of the basilar membrane and the discharges in a single auditory nerve fiber.

time pattern of a sound in the discharge pattern of single auditory nerve fibers. The physiological experiments can only show that such coding exists.

In evaluating these "hearing theories" and their ability to describe the processing the identification of sounds in the auditory system, it should be emphasized that they, like most models of biological systems, have been inspired and greatly influenced by technological developments. That many models of the auditory system are based on spectral analysis no doubt has partly to do with the favor which such methods of analysis have won for describing the performance of numerous man-made systems. In most such applications of spectral analysis, the average energy present in relatively few discrete frequency bands, each usually of a certain relative width, is measured accurately in absolute terms. The basilar membrane, no doubt, performs a kind of spectral analysis, but the output of each channel of the auditory spectral analyzer is not a linear function of input power. Furthermore, the analyzer's elements of output are so numerous that one would rather liken it to a continuous system than to a system with discrete channels. Another difference between the analysis that takes place in the ear and that performed by man-made spectral analyzers is that coding in single nerve cells in the ascending auditory pathway does not emphasize steady-state excitation, but rather emphasizes changes in amplitude and frequency. Thus, the average discharge rate of single auditory nerve fibers in response to a steady pure tone reaches a plateau only 20–30 dB above threshold.

Results of many experiments support the temporal principle rather than the place principle as the basis for pitch perception. Yet, a great deal of experimental and theoretical work has been devoted to spectral analysis of the ear, whereas investigation of temporal analysis has been relatively limited.

At one time, researchers in auditory physiology could be divided into two distinct groups, those who believed that frequency discrimination was accomplished exclusively through the spectral analysis performed according to the place principle, and those who assumed that all frequency analysis was performed through the temporal analysis in the nervous system. Presently, a number of researchers have come to assume that the auditory system uses *both* temporal and spectral analysis in parallel, with more emphasis on the one or the other, depending on the type of sound being studied.

It is thus most likely that the two principles for frequency discrimination are not mutually exclusive, but rather they work in parallel—one playing a dominant role in the analysis of certain types of sounds and the other in the analysis of other types of sounds.

Physiological Measures of the Spectral Selectivity in the Auditory Periphery

Spectral selectivity in the ear is assumed to originate in the cochlea, and most experimental work and hypotheses indicate that it is a result—at least partly if not entirely—of the mechanical properties of the basilar membrane in the cochlea.

Fundamentally different experimental methods have been used to assess the spectral selectivity of the auditory periphery, such as (*a*) measurements of the vibration amplitude of the basilar membrane; (*b*) determination of the threshold of primary auditory nerve fibers in response to pure tones; and (*c*) analysis of the discharge patterns of single primary auditory fibers. Generally, rather different degrees of selectivity have been demonstrated, and great efforts have been devoted to explain these differences. Also, different experimental methods applied to the study of spectral selectivity on the same level have shown widely different results.

Several questions thus emerge when results of past and more recent studies on the auditory nervous system are considered.

1. Is it possible to obtain a *general* picture of the response of the system to complex sounds on the basis of results obtained using simple sounds?
2. Can we draw conclusions about how the vibration amplitude varies *along* the membrane on the basis of the spectral selectivity at different points along the basilar membrane?
3. Can we draw conclusions, on the basis of the firing pattern of single nerve fibers, about the functionally important spectral selectivity to complex sounds by observing the change in the response pattern as the stimulus is varied in frequency and intensity? Can we predict, for example, how the firing pattern in many nerve fibers in the response to a constant sound will manifest itself on the basis of such data?

Recent developments in experimental methods, making it possible to perform intracellular recordings from hair cells, by Russell and Sellick (1978) have contributed valuable information regarding analysis performed in the inner ear. The introduction of statistical signal processing methods for studying discharge patterns of single nerve fibers and nerve cells has also opened new ways to study the function of the peripheral auditory system. This method makes use of stochastic signals (various types of noise) in determination of the frequency selectivity properties of the auditory periphery (de Boer, 1968; Møller, 1977). Recording from many nerve fibers in the same animal responding to the *same* stimulus (pure tones) is another method that provides new information about the spectral selectivity of the ear (Pfeiffer & Molnar, 1970).

It was once believed that lateral inhibition in the auditory nervous system increased the relatively poor frequency selectivity (as measured by von Békésy, 1942) of the basilar membrane. When it became evident that there was no anatomical basis in the inner ear for such sharpening and when physiological experiments showed that spectral selectivity of the auditory periphery is instantaneous in action (Møller, 1970b) and the same at the level of the inner hair cells as in primary auditory nerve fibers (Russell & Sellick 1978) the theory of lateral inhibition was abandoned. It was proposed that types of interactions occur at more peripheral levels, such as among the hair cells. This proposition gave rise to a number of hypotheses that claimed to explain the discrepancy between selectivity based upon measurements of the mechanical motion of the basilar membrane and selectivity based upon the discharge patterns in the auditory nerve fibers or frequency threshold curves (Zwislocki, 1974).

These latter hypotheses were also abandoned when proper experimental results became available. The first successfully performed intracellular recording from hair cells by Russell and Sellick (1977) clearly showed that the same degree of spectral selectivity that exists in primary auditory nerve fibers also exists in the inner hair cells. Preferred hypotheses claim that the high degree of spectral selectivity in inner hair cells is based upon mechanical excitatory patterns of vibration of the basilar membrane (Duifhuis, 1976). (For a comparison of some different sharpening schemes, see Nilsson, 1977.)

One hypothesis regarding the origin of spectral selectivity in auditory periphery put forward by Evans (1975a, b) claims that the spectral selectivity present in the auditory nerve has two fundamentally different origins—namely the mechanical properties of the basilar membrane and a "second filter." The first filter is assumed to have the shape of a lowpass filter with a relatively sharp cutoff, whereas the second filter is a sharp-tipped bandpass filter that is responsible for the sharp tip of the familiar neural tuning curves of single auditory nerve fibers.

SPECTRAL SELECTIVITY OF THE BASILAR MEMBRANE

In Chapter 1, a description was given of how von Békésy (1942), using cadaver ears and a stroboscopic technique, measured the vibration amplitude at various locations on the basilar membrane. These were to be the only direct measurements of this important function of the inner ear for years to come, until 1967 (Johnstone & Boyle, 1967), when the *Mössbauer technique* was used in studies of the motion of a single point on the basilar membrane. Chapter 1 pointed out that subsequent measurements performed by Rhode (1971) showed a greater frequency selectivity than those of von Békésy (1942), whereas other methods indicated that the basilar membrane had a filter function similar to that of a lowpass filter.

There is relative agreement as to the slope of the high-frequency cutoff (skirt) of the basilar membrane's tuning curves, regardless of whether the Mössbauer effect or other methods are used. A nonlinearity, giving rise to a narrowing of the tuning curves as intensity is decreased, has only been demonstrated in squirrel monkeys at the 7 kHz point (Rhode, 1971). In a following section, the basilar membrane frequency selectivity will be compared to the selectivity determined on the basis of recordings from single auditory nerve fibers and psychoacoustic methods. In addition, results of experiments based on the impulse activity in the auditory nerve will be described in more detail than was done in Chapter 2.

Spectral Selectivity at the Level of the Neural Transduction Process

A key experiment, which elucidated the nature of the frequency selectivity of the auditory periphery and increased the understanding of the discrepancy between the acuity of the tuning in primary auditory nerve fibers and the mechanical tuning of the basilar membrane, is that experiment in which intracellular recordings were made from individual hair cells. Such recordings make it possible to determine whether a neural frequency-selective element exists in addition to the mechanical selectivity of the basilar membrane. Thus, Russell and Sellick (1977, 1978) showed that tuning curves based on intracellular recordings from inner hair cells in the guinea pig cochlea in response to pure tones show a sharp tuning that is very similar to that of auditory nerve fibers and the shape of the obtained frequency tuning curves closely resembles that of auditory nerve fiber tuning curves. An ac potential, which corresponds to the cochlear microphonic potential, and a dc potential, which is related to the summating potential, were recorded. Some of their results are shown in Figure 3.3. The curves shown in this figure were obtained by measuring the sound level required to evoke a certain small change in the dc level recorded from the hair cells. The lower curve in Figure 3.3 that represents the sound level required to produce a just-recordable dc change may be regarded as comparable to a frquency threshold curve of a single auditory nerve fiber. The hair cell frequency threshold curve in Figure 3.3 has an extremely sharp tip, indicating a very sharp tuning at the level of the hair cells. When the sound level required to evoke a higher dc voltage is determined (iso-response curves), it follows that the sharpness of the tuning decreases with increased intensity. Such iso-response curves show a nonlinearity, the nature of which is similar to that shown by Rhode (1971, 1978; Rhode & Robles, 1974) for the mechanical tuning of the basilar membrane in the squirrel monkey (see p. 78) and to that shown by Møller (1977) for single auditory nerve fibers. Other properties of hair cell tuning curves seem similar to those of the auditory nerve fibers—tips of the hair

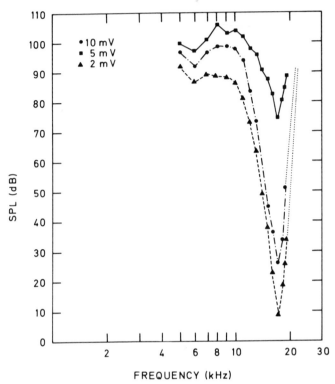

FIGURE 3.3. Results of intracellular recordings from inner hair cells in the guinea pig. The individual curves show the intensity of a pure tone that evoked a certain dc shift of the intracellular response as a function of the frequency of the tone. (From Russell & Sellick, 1977.)

cell tuning curves seem to have the same vulnerability to various noxious agents as do those of tuning curves of the auditory nerve fibers (Evans, 1975b).

Most of the spectral filtering properties that can be measured in single auditory nerve fibers are already established at the level of sensory transduction in the cochlea (hair cells). Also, with two-tone suppression, great similarities were found between responses obtained from intracellular recordings from inner hair cells and responses from single auditory nerve fibers.

Using a pair of pure tones as stimuli, Sellick and Russell (1979) could show that the response evoked by one tone at the cell's characteristic frequency (CF) can be inhibited by a second tone within restricted frequency ranges above and below the cell's response area to one tone. In fact, two-tone suppression shown in the response from inner hair cells is identical to the two-tone suppression shown in primary nerve fibers. It seems, then, that the frequency selective properties of response characteristics of primary auditory nerve fibers may already exist at the intracellular level of the inner hair cells. Also, stimulus response functions relating the magnitude of the dc potential to stimulus intensity are similar to rate functions of auditory nerve fibers (Sachs & Abbas, 1974) and differ in that the neural rate functions have a sigmoidal shape, whereas the intracellular potential rises linearly from the noise level and does not seem to have a threshold. Also, the saturation level of the dc potential is lower for frequencies at and below the cell's characteristic frequency compared with frequencies above—a relationship similar to the way the neural rate function saturates.

Results of recordings from inner hair cells indicate that there is no appreciable electrical interaction between individual inner hair cells, as has been previously suggested (see Manley, 1978).

FREQUENCY SELECTIVITY IN THE AUDITORY NERVE TO BROADBAND STIMULI

Chapter 2 described how single nerve fibers in the auditory nerve responded only to tones within limited frequency ranges. It was also noted that the threshold had its lowest value at a certain frequency, called the fiber's characteristic frequency. It was assumed that this frequency tuning originated, though not necessarily exclusively, in the mechanical tuning of the basilar membrane. In this chapter, the frequency selectivity of auditory nerve fibers will be considered in more detail. The relationship between this neural tuning and the mechanical tuning of the basilar membrane will also be discussed. Frequency selectivity above threshold and the use of stimuli other than pure tones will be considered.

Most of the work done on the spectral selectivity of the auditory analyzer has involved pure tones. Responses over a large frequency range have been measured by presenting many tones with different frequencies and then combining the results. It has been suggested that the results of such experiments could be used for estimating responses to complex sounds containing many spectral components. Since tones are presented one at a time in such experiments, possible interation between the different spectral components of a complex sound cannot be observed. Indeed, recent experiments have produced strong evidence that responses to complex sounds generally *cannot* be estimated on the basis of responses to pure tones presented one at a time. Furthermore, the aim of most of the pure tone experiments has been to determine the thresholds for different frequencies. Such threshold functions do not necessarily reflect the function of the system at physiological sound levels.

Correlation Techniques

The first investigator to use the correlation technique in studies of the spectral resolution of the auditory periphery was de Boer (1967). His method, a special type of correlation techique called *reverse correlation*, was particularly suited for use in connection with the common average-response computer. The correlograms obtained are approximations of cross-correlation functions, and consequently, the frequency domain transfer function can be computed by Fourier-transforming these reverse correlograms. De Boer found a frequency selectivity similar to that shown in the frequency tuning curves obtained using pure tones as stimuli.

Later studies, in which pseudorandom noise was used as the stimulus, showed that the frequency selectivity of responses from primary auditory nerve fibers varied in a systematic way with the intensity of the stimulus (Møller, 1977).

The method using pseudorandom noise makes it possible to obtain the transfer function of the system being tested over a large range of sound intensities. Applied to responses of the auditory nerve, these transfer functions represent the frequency tuning of the peripheral auditory system and thereby yield important information about the spectral resolution of the system when stimulated with complex sounds at levels above threshold. Since these methods are dependent upon the phase-locking of discharges to the waveform of the sound, they are restricted to the low-frequency range of up to about 5 kHz.

Recordings of the discharges in single auditory fibers in the rat, upon stimulation with pseudorandom noise (Møller, 1977), have shown that fibers with a characteristic frequency between .7 and 4 kHz have a narrow transfer function if the stimulus sound has an intensity just about threshold. For noise levels just above threshold, the computed transfer functions are similar in shape to the frequency threshold curves of the fiber (de Boer, 1967, 1968, 1969; Møller, 1978b). When intensity is increased above threshold, the width of these transfer functions increases with stimulus intensity. At the same time, as the width of the transfer function becomes larger, the characteristic frequency shifts downward.

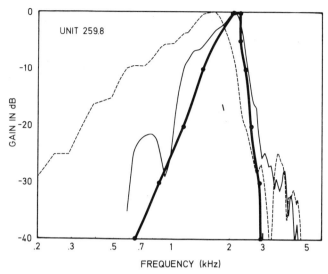

FIGURE 3.4. Frequency selectivity of a single auditory nerve fiber in a rat determined on the basis of the response to noise stimulation (thin lines) and the threshold to stimulation with pure tones (heavy lines). The two curves representing the response to noise were obtained at two different stimulus intensities: solid lines, 15 dB above threshold; and dashed lines, 70 dB above threshold. (From Møller, 1978b.)

Examples of these transfer functions are shown in Figure 3.4 for two different stimulus intensities. The graph also shows the frequency threshold curve of the fiber. The increase in bandwidth of these transfer functions with increase in sound intensities is shown in Figure 3.5. The Q value and the characteristic frequency are shown in Figure 3.6 as functions of intensity. (The Q value is a measure of frequency selectivity. The higher the Q value, the greater the selectivity will be. The Q_{10} value is defined as CF/B where CF is the center frequency and B is the bandwidth measured at 10 dB below maximal response.) As the stimulus intensity is increased, the Q_{10} value decreases. This is a general finding, although most pronounced for fibers with characteristic frequency above 2 kHz. The downward shift in characteristic frequency with increasing stimulus intensity shown in Figure 3.6 is also a general finding, but it is more pronounced for fibers with a higher rather than lower characteristic frequency (Møller, 1977, 1978b). It is somewhat puzzling that other investigators do not see a similar nonlinearity (Wilson & Johnstone, 1975; Evans, 1977).

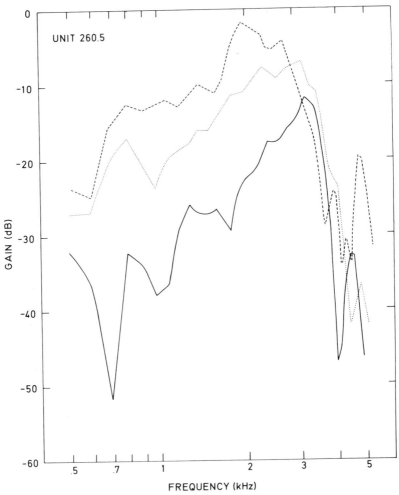

FIGURE 3.5. Frequency selectivity in a single auditory nerve fiber in a rat determined from the responses to pseudorandom noise. The three curves represent different sound intensities, of 42 dB SPL (solid line), 62 dB (dotted line), and 82 dB SPL (dashed line) measured in a ⅓-octave band. The curves were normalized with regard to their peak value and then displaced 5 dB from each other to facilitate comparison. (From Møller, 1978b.)

These results point toward the existence of a nonlinear filter function, where the width of the derived transfer functions increases with stimulus intensity and their characteristic frequency shifts downward. Results obtained from recordings from single nerve fibers using noise as stimuli thus

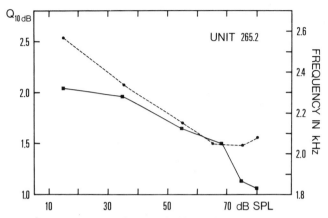

FIGURE 3.6. Center frequency (solid lines and right-hand scale) and Q_{10} (dashed lines and left-hand scale) of selectivity curves such as those shown in Figure 3.5, as a function of the sound intensity (in decibels SPL measured in a ⅓ octave band). (From Møller, 1977.)

support Rhode's (1971) finding that the basilar membrane is nonlinear, in the sense that the acuity of its frequency tuning is a function of the sound intensity. Furthermore, the findings just mentioned indicate that basilar membrane nonlinearity extends over a large part of the physiological intensity range of the auditory system.

When nonlinearities are discussed, it is assumed that there is a concomitant generation of harmonic components when the input (stimulus) is a sinusoidal signal (pure tone). There is a clearly evidenced nonlinearity in the peripheral auditory system that results in generation of distortion products (see p. 154). Nonlinearities in the peripheral auditory system, either by studies of the mechanical motion of the basilar membrane or by the responses of single auditory nerve fibers, may, however, be of a different type. These nonlinerities manifest themselves mainly by a change in the spectral filtering characteristics, such as an increase in the bandwidth with increasing sound level. This is, no doubt, the behavior of a nonlinear system, but it may differ from that of nonlinear systems in which there is a distortion of the waveform and subsequent generation of spectral components that do not exist in the input. The observed nonlinearity of auditory periphery may be described as a change in filtering properties with sound intensity. If this change occurs slowly (much slower than the time for one cycle of the sound), it will not result in any distortion of the waveform at its output, and the technique of linear systems analysis may be used to obtain valid results.

Comb Noise

Another method for determining spectral acuity of the peripheral auditory analyzer entails the use of a noise stimulus—the energy of which varies with frequency in a periodic way. Such a stimulus is produced by adding broadband noise to the same broadband noise in a time-delayed version. Resulting noise is called *comb noise.* Spacing of the peaks and valleys of energy is a function of the time interval between noise and its time-delayed version. Such methods have been used in psychoacoustical (Wilson & Evans, 1971) as well as in neurophysiological studies (Evans & Wilson, 1973; Wilson & Evans, 1971). In neurophysiological experiments, spacing between peaks and valleys is decreased until there is no change in the discharge rate when peaks are exchanged for valleys. Wide spacing between peaks and valleys causes a marked difference in the discharge rate when peaks and valleys are exchanged, whereas narrow spacing brings their discharge rates nearer to one another (Figure 3.7). These stimuli are analogous to the sinusoidal gratings used in studies of the visual system (see, for example, Campbell, Cleland, Cooper, & Enroth-Cugell, 1969).

All of these observations are taken as an indication of the ability of the peripheral auditory analyzer to distinguish fluctuations in the spectrum of a noise, and thus of its ability to perform spectral analysis. The limit for detecting the difference between a flat spectrum and a nonflat spectrum is a measure of frequency selectivity of the peripheral auditory analyzer. On the basis of spacing between valleys and peaks of the comb noise, where a difference can just be detected, it is possible to compute the equivalent bandwidth value of the peripheral spectral analyzer.

In gereral, single auditory nerve fibers have been shown to respond to comb noise in the same way as does a linear bandpass filter that has the same width and shape as frequency tuning curves obtained using conventional techniques and pure tones (Evans, 1975a).

Spectral Integration of Broadband Noise

Integration of energy across frequency of broadband sounds, such as broadband noise, may be used as a measure of the width of spectral analysis filters and thus as a measure of spectral resolution. In a linear filter, energy is integrated over the bandwidth of the filter. A broadly tuned filter transmits more noise energy than a narrow filter does, but the transmission of a pure tone is the same in these two filters. The difference

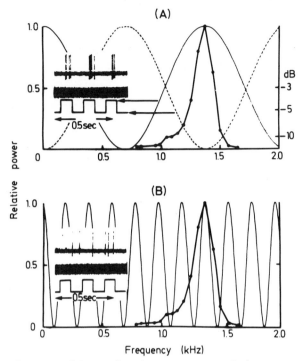

FIGURE 3.7. Illustration of the use of comb noise in assessing the frequency selectivity of single auditory nerve fibers. The upper graph illustrates the response to comb noise with widely spaced peaks, and the lower graph shows responses to noise with narrow-spaced peaks. Filled circles and thick, continuous lines show frequency tuning curves of a single auditory nerve fiber plotted as a bandpass-filter function in relative power and frequency coordinates. (A)Thin, continuous, sinusoidal lines show the envelope of the noise spectrum, which was so adjusted that one of its peaks coincided with the center frequency of the nerve fiber. Note that both the frequency scale and the vertical scale (left) showing noise power are linear scales. (B)Narrowly spaced noise spectrum with only the normal spectrum shown. Inserts show records of spike discharges in response to the comb-filtered noise that was alternated between "normal" and "inverted" every 100 msec. The middle trace shows an oscillogram of the comb-filtered noise. Note that in (A) the discharges are related to the switching of the noise from normal to inverted whereas there is no discernable difference in the discharges to the two versions of the noise in (B). (From Evans & Wilson, 1973.)

in transmission of noise and tone thus becomes a measure of bandwidth. This has been used in studies where recordings of discharge rates of single auditory nerve fibers (Evans, 1975a) and of cells in the cochlear nucleus

(Møller, 1970a) in response to pure tones (at center frequency) and broad-band noise were compared. The relationship between excitation by a pure tone at the characteristic frequency of a nerve fiber and the excitation by a broadband noise is a measure of the spectral integration of nerve fibers. The difference in the threshold for tones and noise, or the difference in the intensity of a tone and of a noise that gives rise to the same excitation, have been used as measures of spectral integration and, ultimately, of the bandwidth of auditory nerve fibers and cells in the cochlear nucleus. Figure 3.8 shows results from threshold measurements in single auditory nerve fibers (Evans, 1975a).

Results obtained using the same method in the cochlear nucleus of the rat showed similar results, despite the fact that they were obtained in the cochlear nucleus and in another species (Møller, 1970a). In these experiments, comparisons similar to those drawn between the width of the frequency threshold curves and the bandwidth values at super-threshold values (10 dB above threshold) were derived from noise measurements. Most of the values obtained from the latter method are similar to those obtained at threshold, but a few results obtained above threshold show a narrower bandwidth on the basis of noise measurements compared with the width of the frequency threshold curves.

These findings have been interpreted as reflecting the influence of inhibitory areas surrounding the excitatory area of these units (Møller, 1970a; Evans & Wilson, 1971, 1973). A conclusion that may be drawn from these experiments is that both primary auditory nerve fibers and most cochlear nucleus units integrate noise over frequency near threshold in a similar way as a linear filter with the bandwidth equal to that of frequency tuning curves would. Above threshold, some units in the cochlear nucleus show a smaller degree of spectral integration in response to noise than expected on the basis of their frequency threshold curves. It is worth noting that auditory nerve fibers, as well as many neurons in the cochlear nucleus, do not respond well (some do not respond at all) to broadband noise above a certain sound level (Greenwood & Goldberg, 1970; Gilbert & Pickles, 1980).

Recordings from Many Nerve Fibers in Response to the Same Tone

A fundamentally different way of determining frequency selectivity of the auditory nerve fibers was developed by Pfeiffer and Kim (1975). It makes use of recordings from a large number of nerve fibers in the

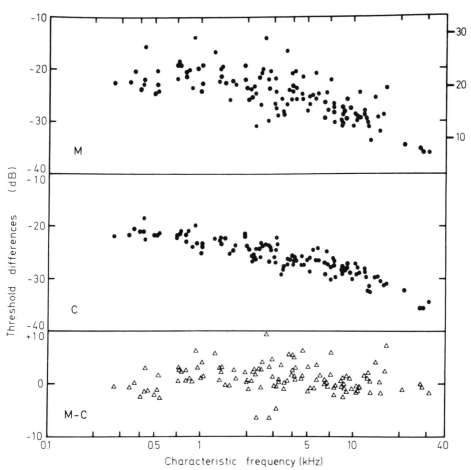

FIGURE 3.8. Bandwidth of auditory nerve fibers estimated on the basis of the threshold to broadband noise and pure tones at center frequency (CF). Difference between noise and tone threshold from 118 cochlear nerve fibers from five cats plotted against the CF of each fiber. M, difference between noise threshold value and the threshold value for a tone at CF. The noise was expressed in RMS power in a 1-Hz-wide band (right-hand ordinate gives the values measured using a full noise bandwidth of 21.5 kHz). In that way the difference expressed in decibels becomes equal to $10 \log_{10}$ of the effective bandwidth if the filter under test functions like a linear filter. C, bandwidth values measured from frequency threshold curves and converted into the same units as in graph M. The bandwidth values used are the "effective" bandwidth, which is the width of a rectangular filter that is equal in area to the frequency threshold curve when plotted on linear power and frequency coordinates. M-C, difference between corresponding measured and computed values. If the filter had been linear, this difference should have been zero. (From Evans, 1975, based on Evans & Wilson, 1971, 1973.)

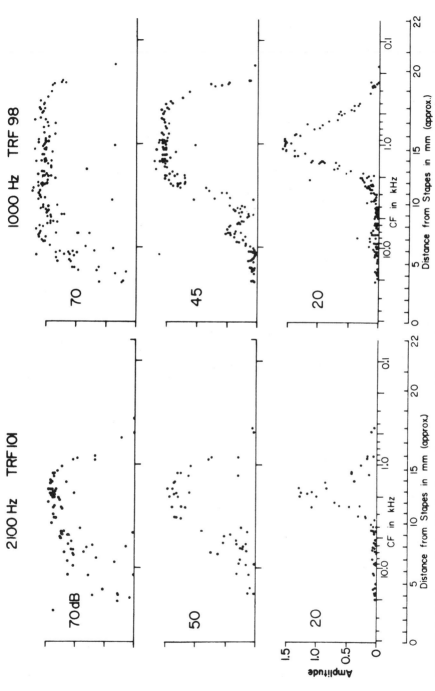

FIGURE 3.9. Relative amplitude of the fundamental component of the response to 2100 Hz and 1000 Hz at three different intensities plotted as a function of the center frequency of the fibers. Each dot represents one fiber. These results were obtained from two anesthetized cats. The horizontal axis gives the center frequency of each of the individual fibers studied and the estimated distance from the stapes. Each column is from a different animal. (From Pfeiffer & Kim, 1975.)

210

auditory nerve of an experimental animal responding to the same (one or two) tones. Pfeiffer and Kim obtained period histograms of the recorded discharge pattern locked to the periodicity of stimulus tones. Due to phase-locking of the discharges to the sound waveform, these histograms became modulated with the frequency of the stimulus and its harmonics. The degree of modulation of the histograms is assumed to be a measure of the degree of excitation that a certain stimulus gave rise to. Each fiber was identified by its characteristics frequency. By Fourier-transforming these histograms, amplitude and phase of the modulation of the discharge density could be determined. When recordings from a large number of fibers are compiled, it is possible to establish a measure of how excitation varies along the basilar membrane for a single tone at different intensities. This approach is fundamentally different from establishing the frequency selectivity of a single nerve fiber or a single point on the basilar membrane. The method developed by Pfeiffer and Kim may provide a more physiological measure of the frequency selectivity of the ear. Figure 3.9 shows examples of results for two tones at three different sound intensities. These results comprise recordings for 240 single nerve fibers. The graphs in Figure 3.9 show the relative amplitudes of modulations in histograms of discharges from nerve fibers of different characteristic frequencies. The results are plotted as a function of characteristic frequency and are in accordance with the assumed distribution of nerve fibers along the basilar membrane. Each graph in Figure 3.9 represents one sound stimulus level and each dot represents one fiber. Figure 3.10 shows the amplitude and the phase angle of the fundamental component of modulation of response to stimulation with seven different frequencies obtained at a level of 20 dB. There is a pronounced frequency selectivity to the weak tones, but as the intensity is raised the excitation spreads along the basilar membrane and the excitation pattern reaches a plateau showing almost equal level of excitation of a large part of the basilar membrane when the sound intensity is raised to physiological levels.

Bandwidth Values of Cochlear and Neural Tuning as Measures of Frequency Selectivity: Do We Measure Selectivity Correctly?

Much experimental work on the function of the auditory system has been concerned with determining the bandwidth of frequency tuning in the peripheral ear. That is a direct result of the hypothesis that the

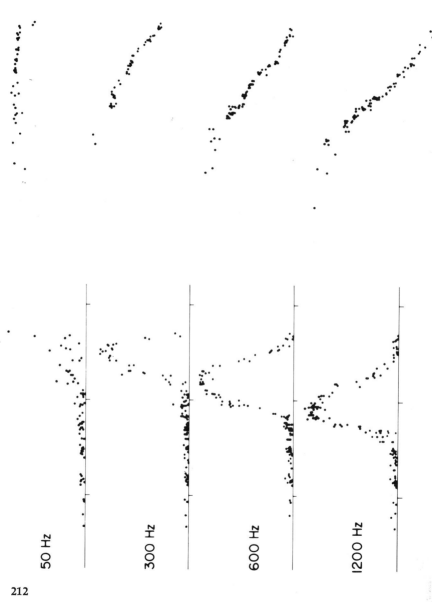

50 Hz

300 Hz

600 Hz

1200 Hz

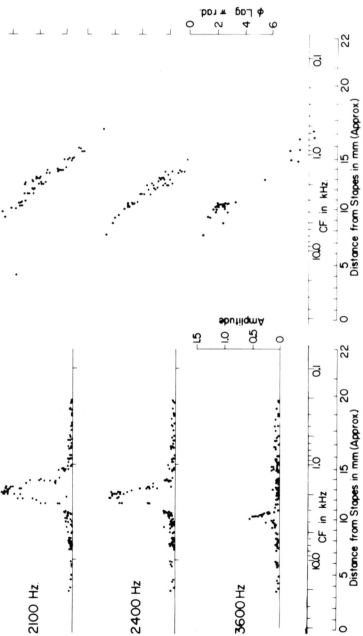

FIGURE 3.10. Amplitude and phase of the fundamental frequency of the modulation of the responses from 240 nerve fibers in response to seven different frequencies obtained in a single animal. The stimulus level was 20 dB. The phase angles shown are relative to the phase angle of the sound at the eardrum. (From Pfeiffer & Kim, 1975.)

peripheral auditory system is comparable to a series of bandpass filters. A further consequence of this hypothesis has been that the frequency selectivity of the auditory periphery has generally been described in the same terms as the frequency selectivity of man-made bandpass filters—namely by a single value of bandwidth—measured as the difference between the frequencies at which the threshold is 3 or 10 dB higher than it is at the fiber's characteristic frequency. This way of describing the tuning characteristics may be adequate for conventional man-made bandpass filters that are linear and have rather steep skirts on both sides of the center frequency. The width of such a filter at the point where the attenuation is 3 dB (half-energy) below the peak transmission provides a reasonably accurate description of its frequency selectivity. Since the selectivity curves of the basilar membrane motion, as well as the frequency threshold curves of single auditory nerve fibers, are highly asymmetrical and have high frequency slopes that are much steeper than the low frequency slopes, bandwidth values will be almost entirely dependent on the low frequency slopes.

The slope of the low frequency part of the selectivity curve of the basilar membrane measured using the Mössbauer effect or the capacitive probe, is derived to a great extent from the frequency characteristics of the middle ear. The transfer function of the basilar membrane is approximately a lowpass function. Since the high-frequency part of the selectivity curve is much steeper than the low-frequency part, the bandwidth values obtained are mainly a function of the slope of the low-frequency part of the transfer function, thus chiefly measuring the function of the middle ear and depending on cochlear filtering, only to a much lesser extent.

It is, therefore, doubtful that the characteristics of such spectral filters as those of the auditory periphery can be adequately described by a single bandwidth value. This manner of description is certainly oversimplified and should be viewed with caution.

Determination of Auditory Frequency Selectivity Using Psychoacoustical Methods

It has generally been difficult to compare psychoacoustical data with neurophysiological data. Neurophysiological data reflect the function of single elements on a certain level of the ascending auditory pathway. Psychoacoustical data, on the other hand, refer to the entire auditory system as well as to analytical parts of the central nervous system about which we know very little.

PSYCHOACOUSTIC TUNING CURVES

Psychophysical tuning curves are based on the assumption that a tone with an intensity of only a few decibels above threshold will excite only a few hair cells. Determination of its masked threshold may, therefore, be assumed to be analogous to the determination of the threshold of a single nerve fiber of the auditory nerve frequency threshold curves (Small, 1959; Vogten, 1974; Zwicker, 1974; Dallos & Cheatham, 1976; Houtgast, 1973). These two measures of auditory frequency selectivity cannot, however, be directly compared on a quantitative basis, and a direct comparison of the bandwidth of frequency threshold curves (FTC) of single auditory fibers and psychoacoustical tuning curves obtained in the same animals show different values of bandwidth.

The method used to obtain psychoacoustic tuning curves is the only method that provides information about the frequency selectivity of the human ear. Since this selectivity can be directly compared with results of electrophysiological recordings from single auditory nerve fibers in animals, this method has lately been applied to the study of various types of pathological conditions in humans (Zwicker, & Schorn, 1978). Although results have generally been difficult to interpret, it emerges that tips of the obtained frequency threshold curves are affected by a variety of different traumas to the inner ear.

CRITICAL BAND

Zwicker (1954) described the critical bandwidth concept that has evolved from psychophysical data on auditory frequency selectivity. *Critical bandwidth* is the bandwidth within which different spectral components of a sound are added and interact in various ways. Many complex sounds have been reported to change abruptly in their subjective appearance when their bandwidth exceeds the critical bandwidth. Critical bandwidth also defines the limits of the ear's ability to separate individual components of a complex sound. However, the critical bandwidth mechanism is not necessarily associated with a filter in the auditory system, nor is it certain that the auditory diescrimination of spectrum can be described adequately by the bandwidth value of an analogous filter.

Impulse Response Function of the Peripheral Analyzer

Response of a narrow bandpass filter to a brief, transient sound (e.g., a click) is a damped oscillation. Since the basilar membrane has the characteristics of a relatively narrow bandpass filter, it can be expected that its response to click sounds will show a damped oscillation. In a linear system, the impulse response function may also be determined on the basis of the frequency domain transfer function. The complexity and possible nonlinear function of the auditory periphery make it unlikely that such an indirect determination of the impulse response function would become a correct measure of its transfer function. It was mentioned earlier that the response of the basilar membrane to other stimuli cannot always be predicted on the basis of the response to simple sounds, as is the case in a linear system, where the response to any signal can be predicted on the basis of knowledge about the system's frequency domain transfer function or its impulse response function. The impulse response of the basilar membrane was obtained by Fourier transformation of von Békésy's (1943) data by Flanagan (1962). These data are based on the assumption that the basilar membrane is a linear system. It can therefore not be taken for granted that they reflect the real response of the system to a brief transient. It is, therefore, of importance to obtain the impulse response function of the peripheral auditory system directly. Comparison of the impulse response functions obtained at different levels of the auditory system (basilar membrane, hair cells, auditory nerve) provides information about the transformation of the signal that occurs in the different parts of the auditory system.

DETERMINATION OF THE IMPULSE RESPONSE FUNCTION
OF THE BASILAR MEMBRANE

The impulse response function of the basilar membrane has been directly determined as a response to a click sound (von Békésy, 1943; Rhode & Robles, 1974; Robels, Rhode, & Geisler, 1976), and has been shown to be a damped oscillation. This was demonstrated in Chapter 1 (Figure 1.64), where the impulse response of the basilar membrane was determined upon stimulation with short click sounds. The velocity of a point on the basilar membrane was shown as it appears from the Mössbauer measurements upon stimulation with brief click sounds. It is clear that the late parts of the response do not grow as fast with increasing stimulus intensity as the early parts do. This indicates that the basilar

membrane motion has nonlinear properties similar to Rhode's (1971) findings in experiments using pure tones as stimuli. It must be pointed out that
the method employing the Mössbauer effect offers a relatively limited
dynamic range of recordings and that the impulse responses shown in
Figure 1.64 are a result of computations, including approximation of the
response to sinusoidal waveforms. These circumstances, together with
the surgical preparation, and the placement of the radioactive source on
the basilar membrane, introduce elements of uncertainty into measurements of the impulse response function.

Since these recordings of the impulse response function of the basilar
membrane motion are the only ones of their kind, their vaildity cannot be
assessed in the light of other data.

It should be pointed out that the parmeter being measured is the up-
and-down movement (amplitude or velocity) of the basilar membrane. It
is not known which type of movement is the most efficient in exciting the
hair cells. Furthermore, these mechanical measurements are made after
the cochlear capsule has been opened. The specific damage or alteration
caused by this procedure is not yet known.

NATURE OF THE IMPULSE RESPONSE FUNCTION AT THE AUDITORY NERVE

Responses of the single auditory nerve fibers indicate that the response
of the auditory peripheral analyzer to a short click is a damped oscillation. Thus, the peristimulus-time (PST; also known as poststimulus
histogram) histogram of the responses of single auditory nerve fibers
with a low characteristic frequency to click stimulation show distinct
peaks (Kiang *et al.*, 1965; Figure 3.11). The interval between these peaks is
equal to the inverse of the characteristic frequency of the particular fiber.
It is believed that the periodic nature of histograms is a result of the
damped oscillation of motion of the basilar membrane.

Since hair cells are excitated only when they are deflected in one direction, these histograms represent the half-wave rectified version of the motion of the basilar membrane. By obtaining histograms of both polarities
of the responses to the click stimulation, and combining these two
histograms into a compound histogram, one can arrive at a more realistic
idea of the damped oscillatory response of the auditory analyzer (Goblick
& Pfeiffer, 1968; Pfeiffer & Kim, 1973). Because of the influence of the
refractory properties of the neurons and the unknown, nonlinear relationship between discharge rate and excitation, such histograms do not lend
themselves to quantitivative determinations of the amplitude of in-

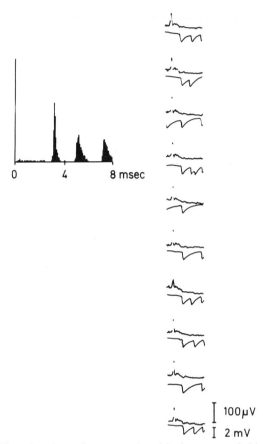

FIGURE 3.11. The right column shows examples of discharges recorded from the auditory nerve in response to click stimulation. The upper traces show discharges from a single nerve fiber and the lower trace shows the gross nerve response recorded from the round window. The graph to the left is a poststimulus-time histogram (PST) of the discharges of the single auditory fiber. The center frequency of the fiber was 540 Hz. (From Kiang *et al.*, 1965.)

dividual waves of the damped oscillation of the auditory analyzer. Therefore, other and better ways of estimating the impulse response function have been proposed. One method, which makes it possible to obtain estimates of impulse response functions of the auditory peripheral analyzer, uses statistical signal analysis of the discharge pattern of single auditory nerve fibers in response to stimulation with pseudorandom noise. The peripheral auditory analyzer is, in general, nonlinear but it may be assumed to be composed of a linear part and a nonlinear part. This separation in a nonlinear and linear part may not necessarily have an anatomical correlate but it is a concept that is valuable in analysis of the function of the peripheral auditory analyzer. The cross-covariance functions between period histograms of the discharges of single auditory nerve fibers and one period of the noise are assumed to be valid estimates of the impulse response functions of the linear part of the peripheral auditory system (Møller, 1977, 1978a).

Typical cross-correlation functions are shown in Figure 3.12. They are clearly seen to be damped oscillations, the duration of which decreases with increasing sound intensity. This indicates that bandwidth increases

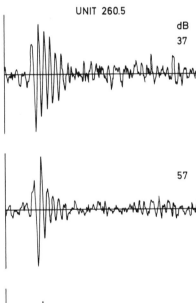

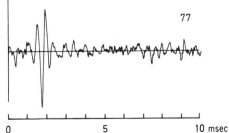

FIGURE 3.12. Estimates of the impulse response functions of a single auditory nerve fiber in the rat for three different sound intensities. The results were obtained from the response of a single auditory nerve fiber in a rat to stimulation with pseudorandom noise. The curves show the cross correlation between the histogram of the responses and one period of the pseudorandom noise. (From Møller, 1977.)

with increasing intensity (cf. p. 78). The procedure used in obtaining the illustrated results is described by Møller (1977).

At first glance, the derived estimates of the impulse response functions in Figure 3.12 resemble the impulse response functions of an ordinary bandpass filter, except for the fact that they change with sound intensity. Closer examination reveals another difference between these impulse response functions and impulse response functions of ordinary bandpass filters. The frequency of the damped oscillations changes along the time axis. This becomes evident by observing the intervals between zero crossings of the individual waves of the damped oscillation, which are not the same in the beginning as at the end of the oscillation (Møller & Nilsson,

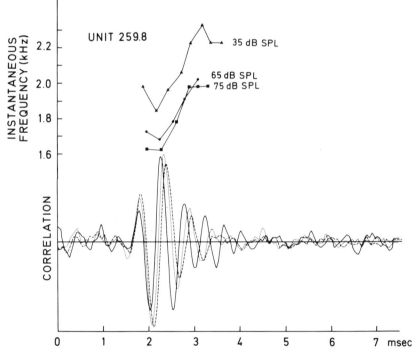

FIGURE 3.13. Estimates of impulse response functions similar to those shown in Figure 3.12 but superimposed to show change in frequency of the damped oscillation. Upper curves show instantaneous frequency. Solid line, 35 dB; dotted line, 65 dB; dashed line, 75 dB SPL (measured in ⅓-octave band). Inverse of the time intervals between zero crossings as a function of time. (From Møller & Nilsson, 1979.)

1979). This is illustrated in Figure 3.13, where typical cross-correlation functions for three different sound intensities are shown, together with the instantaneous frequency, as a function of time. The reason for this frequency modulation lies in the traveling wave nature of the basilar membrane motion. The functional implication of frequency modulation is, among other things, that the motion of the basilar membrane *cannot* be correctly modelled by ordinary filters made up of lumped components or by simulation of such filters. This means that not even the filter characteristics of a single point of the basilar membrane are adequately described by such models, and only models containing or simulating distributed constants are adequate. These matters have been studied using mathematical models and lie outside the scope of this book. Readers are referred to Hall (1980) and Zwislocki (1980) for further discussion of this topic.

Another method involving broadband stimuli, in the determination of the spectral resolution of the auditory analyzer, entails the presentation of paired clicks with a varying interval between the clicks in each pair (Møller, 1970b). This method is based on the assumption that each click evokes a damped oscillatory response along the basilar membrane. The second click is presented before the damped oscillation, evoked by the first click, has died out. Varying the interval between clicks will give rise to peaks and valleys in the total energy of the damped oscillation, depending on whether the second damped oscillation appears in or out of phase with the first (summation versus subtraction). Amplitude of these peaks and valleys will depend on how much the two damped oscillations overlap. Thus, amplitude of oscillations will decrease as the interval between the two clicks in increased and the rate of this decrease is a direct function of the duration of oscillation evoked by each click. Consequently, the duration of the damped oscillation is reflected quantitatively in the number of peaks and valleys that can be detected. The average neural response to two clicks presented with a certain time delay will therefore depend on how the two damped waves evoked by each of the clicks are related with regard to phase and on how much the first damped oscillation has decreased in amplitude before the second click is applied. The total energy of excitation represents the sum of two damped oscillations. Since the total energy of stimulation is reflected in the average discharge rate of single auditory nerve fibers or nerve cells in the cochlear nucleus, this can be measured with relative ease.

Results of experiments carried out in the cochlear nucleus are shown in

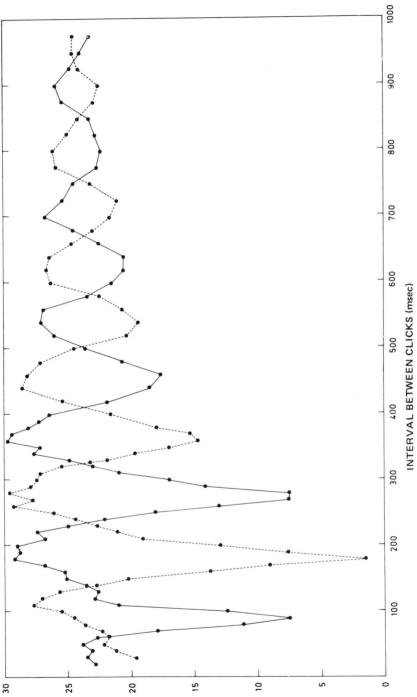

FIGURE 3.14. Response of a nerve cell in the cochlear nucleus of a rat to paired click sounds. The average number of discharges evoked by each click pair has been converted to show equivalent stimulus intensity (dB) and is shown as a function of the interval between the clicks for clicks of the same polarity (solid lines) and opposite polarity (dashed lines). The center frequency of the unit was 5.4 kHz. (From Møller, 1970b.)

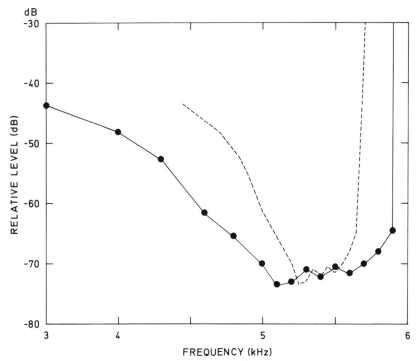

FIGURE 3.15. Frequency threshold curve for a nerve cell in the cochlear nucleus of a rat (solid line) compared with the frequency tuning determined on the basis of the responses to paired clicks (dashed lines). (From Møller, 1970b.)

Figure 3.14. The average number of discharges evoked by a click pair (in equivalent stimulus intensity) is shown as a function of the time interval between the clicks. The amplitude of peaks and valleys decreases as the interval is increased. The response to clicks of opposite polarity produces a similar, but shifted, pattern one-half wave along the horizontal axis. Units with a high center frequency may also be studied by this method, because it is the average excitation level that is measured. Mathematical operations allow for determination of the auto-correlation function of the impulse response of the system on the basis of the results in Figure 3.14. That, in turn, allows determination of the frequency domain (power) transfer function (Figure 3.15). In fact, the tuning curves of cochlear nucleus units determined using paired click sounds are somewhat narrower than the frequency threshold curves determined in the conven-

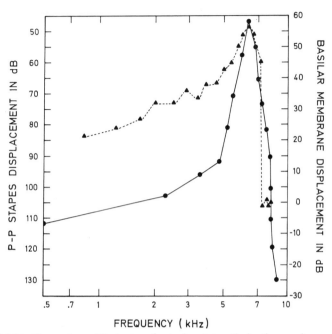

FIGURE 3.16. Comparison of the tuning of a single point on the basilar membrane and that of an auditory nerve fiber. The tuning of the basilar membrane was measured in a squirrel monkey using the Mössbauer effect and pure tone stimulation at 75 dB SPL. (Data from Rhode, 1971.) The tuning in the nerve fiber is represented by a frequency threshold curve from a cat showing the sound intensity required to get a measurable increase in discharge rate. (Data from Kiang *et al.*, 1967.)

tional way using pure tones. This is assumed to be due to a certain degree of convergence of primary fibers upon the cochlear nucleus units. Also, note that frequency threshold curves of cochlear nucleus units are about twice as wide as those of primary auditory fibers (cf. Figure 2.15).

Origin of Spectral Tuning in the Auditory Periphery

It has generally been accepted that the mechanical frequency selectivity of the basilar membrane is the basis for frequency tuning in primary auditory fibers and, consequently, for frequency tuning throughout the auditory nervous system. Using information about the frequency selectivity of the basilar membrane obtained by von Békésy (1942), most investigators, however, assumed that mechanical tuning was sharpened through a neural mechanism of unknown origin. Theories and hypotheses have been put forward to explain the discrepancy between the acuity of the basilar membrane frequency tuning and the frequency (spectral) acuity of the ear as determined first using psychoacoustic methods and, later, using the width of frequency tuning curves of single auditory nerve fibers. As neural tuning curves (frequency threshold curves) gradually became available, a direct comparison of frequency tuning could be made, whereby mechanical tuning was seen to be considerably broader than neural tuning.

The apparent difference between acuity of the mechanical tuning of the basilar membrane and of single primary auditory nerve fibers (Fig. 3.16) further supported hypotheses that an additional frequency selectivity mechanism exists in the peripheral auditory system. This presumed element of sharpening has been referred to as the "second filter" (see Evans, 1975a, b). Without anatomical or physiological evidence, it has been assumed to be located in the vicinity of hair cells and to have bandpass characteristics that are very similar to those of a linear filter (Evans, 1975 a, b). However, most of these theories have been abandoned when techniques of experimentation reached the level where hypotheses could be tested in psychoacoustic and physiological experiments.

Thus, anatomical data have not been able to show any neural structure that could increase selectivity of the basilar membrane. Neurophysiological experiments have disclosed that the full selectivity of the neural tuning *is* available *immediately*, even at frequencies in the 30 kHz region (Møller, 1970b; see p. 221). Sharpening through a neural inhibition mechanism similar to the one achieved in vision, then, is improbable.

Many other suggestions have been put forth about the origin of the sharpening. For example, the discrepancy between neural and basilar membrane tuning curves has been said to result from the fact that von Békésy (1942) used extremely high sound intensities. A nonlinearity in the basilar membrane would naturally result in broader tuning at high sound levels than at low sound levels. Finally, the fact that measurements were performed in cadaver ears could have also played a substantial role in the discrepancy between neural and basilar membrane tuning curves.

Most earlier work on cochlear selectivity has been concerned with the motion of the basilar membrane and the threshold of auditory nerve fibers. Recently, it has become evident that the transformation occurring between motion of the basilar membrane and deflection of the stereocilia may play an important role in describing the discrepancy in tuning of the basilar membrane motion and the tuning in auditory nerve fibers when expressed as frequency threshold curves. Cochlear micromechanics have been given some experimental and theoretical attention—in particular, they indicate that the shear motion between the tectorial membrane and the reticular lamina (Zwislocki & Kletsky, 1980) may play an important role in cochlear spectral filtering and may explain the discrepancy in tuning acuity.

The finding that intracellular potentials of inner hair cells show the same high-frequency selectivity as auditory nerve fibers do (Russell & Sellick, 1977) has put an end to discussion about neural sharpening. Results show that full selectivity exists at the level of hair cells, and, in fact, argue against sharpening of the frequency selectivity by neural interaction. The source of the discrepancy between mechanical tuning of the basilar membrane and the auditory nerve fibers must therefore be sought at, or peripheral to, the hair cells.

A theory put forward by Zwislocki (1974) states that an interaction occurs between inner and outer hair cells by virtue of which sensitivity of fibers terminating on inner hair cells is controlled by outer hair cells. According to this theory, outer hair cells would be likely to reduce the sensitivity of inner hair cells, mainly at frequencies below the characteristic frequency, thus steepening the low-frequency skirt of the frequency tuning curve.

This theory was abandoned (Zwislocki & Kletsky, 1978) due to the experimental results of Russell and Sellick (1977), and it was proposed that mechanical properties of the tectorial membrane and its connection with hairs of hair cells may be responsible for the sharpening (Zwislocki & Kletsky, 1978). Tips of hairs of outer hair cells are assumed to be affixed to the tectorial membrane (Lim, 1972). An up-and-down motion of the organ of Corti is assumed to result in a sliding motion between the tectorial membrane and the reticular lamina, which in turn causes a bending of the hairs (Dallos, 1973a, p. 198) The cilia of inner hair cells do not appear to be in contact with the tectorial membrane, for which reason involvement of this membrane in the excitation of inner hair cells seems highly unlikely. Viscous forces activated by the flow of endolymph surrounding hair cells have been postulated to excite inner hair cells (Billone, 1972; Duifhuis, 1976).

Data showing that hair cells contain actin may offer an explanation for nonlinearity and intensity-dependent frequency selectivity of the basilar membrane (Flock & Cheng, 1977; Flock, 1980). This finding suggests that the coupling between basilar and tectorial membranes can change as a result of physiological processes and that the influence of the tectorial membrane on the mechanical properties of the basilar membrane may vary with sound intensity.

The fact that stereocilia are stiff (Hudspeth & Corey, 1977) suggests that they may act as resonating reeds and in that way contribute to the frequency selectivity of the cochlea (Zwislocki, 1980). The fact that the length of the stereocilia varies along the cochlear partition (Lim, 1980) has strengthened this suggestion.

It is interesting to note that Peake and Ling (1980) have found that single nerve fibers of the alligator lizard have frequency selectivities that are similar to those shown by mammalian cochlear nerve fibers. Two groups of tuning curves could be discerned—one for relatively low frequencies (up to about 1 kHz) and one for higher frequencies. The shapes of low frequency tuning curves differ from those of the higher frequencies in that the former have a steep high-frequency slope, whereas the latter are more symmetrical. Since the vibration of the lizard's basilar membrane (basilar papilla) has no frequency selectivity, it may appear that this animal exclusively has a frequency selectivity that has no relation to traveling waves. It has been hypothesized that frequency selectivity is related to the length of hairs, which varies along the basilar papilla. Since hair cells in the mammalian cochlea are similar to those of the lizard, these results indicate that there may be frequency selectivity in the mammalian cochlea that does not depend on the traveling wave properties of the basilar membrane motion. There are those who assume that this may act to sharpen the frequency selectivity of the basilar membrane.

Other hypotheses about sharpening of frequency tuning in the mammalian cochlea do not involve neural inhibition. Thus, one early hypothesis regarding sharpening is related to basilar membrane displacement itself. This hypothesis is based on von Békésy's (1953a, b) visual observation in the guinea pig cochlea that stimulation of hair cells vibrate in three different directions in the vicinity of the place on the basilar membrane where the transverse displacement was maximal. At the point of maximal displacement, cells vibrate up and down. Toward the base of the cochlea, hair cells vibrate radially, and on the apical side of the location of maximal vibration, hair cells vibrate longitudinally. Von Békésy (1951) furthermore showed that cochlear microphonics were produced only by radial displacement of a hair cell. Spatial distribution of "shear waves" along the basilar membrane has been estimated from computer simulation (Khanna, Sears, & Tonndorf, 1968). Several studies indicate that shear forces are more efficient in exciting hair cells and have a greater frequency selectivity than the up-and-down motion of the basilar membrane. Computer simulations have shown that the longitudinal shear wave is much more localized and the proximal slope is steeper than the traveling wave illustrated in Figure 3.17 (for more details, see Dallos, 1973a).

Much attention has been devoted to functional effects of differentiation of hair cells in the mammalian cochlea into inner and outer hair cells. An earlier hypothesis stating that the outer hair cells had a greater sensitivity than the inner proved unlikely when Spoendlin (1970) found that only about 5% of the fibers of the auditory nerve innervate outer hair cells, whereas 95% of the fibers innervate inner hair cells.

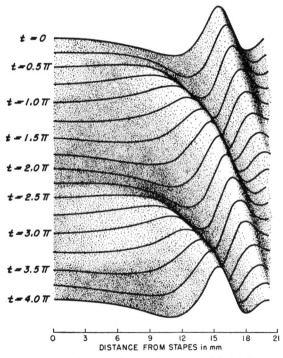

FIGURE 3.17. Computer simulation of the traveling waves of the basilar membrane. Instantaneous displacement is shown. Solid lines represent displacements at $\frac{1}{8}$ cyle intervals. (From Khanna et al., 1968.)

Others (Dallos, Billone, Durrant, Wang, & Raynor, 1972; Dallos, 1973a, b; Dallos & Cheatham, 1976; Dallos & Harris, 1978) however, have taken the view that outer hair cells increase the sensitivity of inner hair cells. After destroying outer hair cells in guinea pigs by adminstration of ototoxic antibiotics (kanamycin), Dallos et al. (1972, 1973a, b, 1976) observed a general increase in the behavioral threshold. Results were interpreted showing that, in the absence of outer hair cells, inner hair cells cannot function at low intensities, but they can function almost normally at high intensities. The well-known clinical phenomenon of loudness recruitment in man—by which people with impaired hearing perceive sounds that are above their elevated hearing thresholds as if these sounds had the same subjective loudness they would have if perceived by people with normal hearing—resembles these findings in Dallos' animal experiments.

Dallos' results also indicate, at least in certain parts of the basilar membrane, that inner hair cells are excited by the vibration velocity of the basilar membrane, whereas outer hair cells seem to be excited by the linear motion of the basilar membrane. As described in Chapter 2, Sellick and Russell (1980) showed that inner hair cells respond to basilar membrane velocity only for frequencies below 100–200 Hz; above that they seem to respond to basilar membrane displacement. These results were obtained in intracellular recordings from inner hair cells in guinea pigs.

Further indication of the complexity of the neural transduction process in the inner ear comes from studies by Pierson and Møller (1980a, b), who found evidence for the existence of two populations of hair cells. Intracochlear recordings from guineau pigs after various types of manipulation of the cochlea, such as anoxia and displacement of the basilar membrane, were different from those recorded before manipulation, and indicated that the cochlear microphonic potential is generated by two populations of hair cells, one of them being more sensitive to anoxia than the other. Superimposing a low-frequency sinusoid on the test sound resulted in a modulation of the amplitude of the recorded cochlear microphonic that was assumed to be a result of a difference in sensitivity at different degrees of deflection of the basilar membrane. Modulation of cochlear microphonics being sinusoidal with a frequency equal to the superimposed low-frequency tone could be reversed in polarity by anoxia (Pierson & Møller, 1980b).

Experimental work shows a number of possibilities for interaction between hair cells or populations of hair cells that may contribute to the frequency selectivity of the inner ear.

Functional Implications of Basilar Membrane Selectivity

When considering functional implications of basilar membrane selectivity, it should be kept in mind that it is probably the distribution of vibration amplitude *along* the basilar membrane that is functionally important. However, physiological studies of frequency selectivity, with a few exceptions, focus on the change in vibration amplitude of a single point on the basilar membrane or on the change in threshold of a single nerve fiber when the frequency of a stimulus tone is varied. This seemingly unphysiological way of studying frequency selectivity in the auditory periphery is a result of the fact that no experimental method that allows observation of the vibration along the membrane, is yet available, nor is there any method available that allows observation of the discharge pattern in a large number of nerve fibers at the same time. The measurement of vibration at single points on the membrane and recording of the impulse activity in one single nerve fiber at any time are thus the only methods available for studying frequency selectivity in the inner ear.

With our knowledge about the function of the basilar membrane, it is not possible to deduce the distribution of vibration along the membrane on the basis of results acquired from isolated points on the membrane. The distribution of vibration amplitude along the basilar membrane depends on factors that have little or no influence on the vibration amplitude of a single point in response to a pure tone. Thus, the change in propagation velocity of the wave motion along the basilar membrane is of fundamental importance to the distribition of vibration amplitudes along it, whereas the vibratory amplitude of a single point on the membrane as a function of frequency is virtually independent of the velocity of wave propagation. If the propagation velocity of the traveling wave along the membrane markedly decreases as the wave travels toward the helicotrema, its amplitude is likely to increase because the same or nearly the same energy is compressed within a smaller section of the membrane. This is counteracted by frictional losses along the membrane, which tend to decrease the amplitude of the traveling wave by the conversion of kinetic energy into heat. It has been pointed out that frictional losses along the basilar membrane may increase from the base towards the helicotrema, but since various studies report different values, precise knowledge about the magnitude of the changes in the losses along the basilar membrane is lacking.

The distribution of the traveling wave's vibratory amplitude along the basilar membrane when the ear is stimulated with a pure tone is not known in detail. In fact, some widely accepted models of basilar membrane vibration describe the amplitude of the traveling wave envelope as almost constant up to that point on the membrane where it decreases, whereas other models suggest that the wave amplitude increases en route from the base of the cochlea toward the point of decrease.

One of the few attempts to visualize the distribution of vibratory amplitude experimentally along the basilar membrane has been made by Pfeiffer and Kim (1975). Neural data shown in Figure 3.10 represent a sharper, and more acute frequency selectivity than the mechanical tuning of the basilar membrane shown by von Békésy (1942) and Rhode (1971). It is also important to mention that the results in Figure 3.10 were obtained at a level of 20 dB, whereas Rhodes (1971) were obtained at 70–90 dB and von Békésy's (1942) were obtained at a sound level of about 145 dB. On the basis of the experimental data shown in Figure 3.10, Pfeiffer and Kim (1975) arrived at a synthesis of the envelope of the traveling wave along the basilar membrane (Figure 3.18).

The traveling wave envelope increases in amplitude as it travels from the basal part of the membrane toward the apex, reaching a prominent peak at a certain point, and decreasing rapidly beyond that point. There is also a widening of the wave envelope with increasing sound intensity; thus, the tuning is sharper at 20 dB SPL than at higher sound levels (see Figure 3.9). In fact at the 70 dB level there is little tuning. This means that at physiological sound levels, there is no significant degree of frequency selectivity (tuning) in the auditory periphery. These results seem to agree with results obtained using broadband noise to reflect the tuning of a single auditory nerve fiber (Møller, see p. 218). They also show a significant reduction of frequency selectivity with increasing sound level.

Vowel-like sounds have been used in studies of the frequency selectivity of the auditory periphery. Such sounds have a broad spectrum with a number of peaks (formants), which have become the subject of neurophysiological experiments similar to those performed by Pfeiffer and Kim (1975). These studies have tried determining to what extent the auditory periphery can resolve peaks in the neural discharges of single auditory nerve fibers (Sachs & Young, 1979). Frequency locations of formants characterize different vowels, and it has been claimed that one vowel is discriminated from another on the basis of the frequency location of these spectral peaks (formants).

Results of the experiments by Sachs and Young (1979) show, in short, that the firing pattern of a large population of auditory nerve fibers reflects the spectral pattern of the sound at threshold sound intensities,

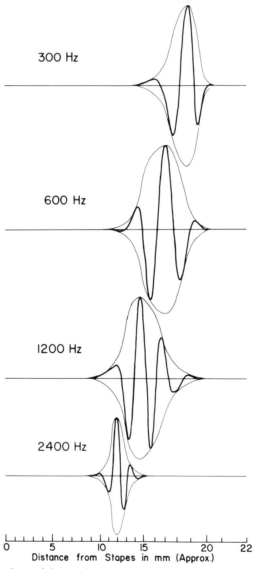

FIGURE 3.18. Synthesis of the traveling wave on the basilar membrane on the basis of the experimental data shown in Figure 3.10. (From Pfeiffer & Kim, 1975.)

and that characteristic peaks (formants) may be discerned clearly. Above threshold, however, peaks in the response pattern become broader for pure tones in accordance with results obtained by Pfeiffer and Kim (1975). At physiological sound levels, vowel sounds resulted in a response pattern which had no peaks that correlated with peaks (formants) in the stimulus spectrum.

These results lead one to doubt seriously that steady-state vowels are discriminated on the basis of their steady-state spectra. Similar studies show that the temporal pattern of discharges in response to vowel sounds carries important information (Sachs & Young, 1980). Also, natural vowels are not steady-state sounds such as those used in the experiments just mentioned. There are large differences in responses to time-varying sounds compared to those of steady-state sounds, when we study neurons in the cochlear nucleus (see Chapter 4).

Temporal Analysis in the Auditory System

Coding of the temporal pattern of a sound in the peripheral auditory system has not been studied as extensively as spectral coding. Physiological studies on temporal coding of frequency have been mainly concerned with whether or not the neural discharge pattern of single auditory nerve fibers is modulated in accordance with the temporal pattern of pure tones and other simple sounds. It has been demonstrated in studies in various animals that discharges of primary auditory fibers are more or less phase-locked to waveforms of periodic low-frequency sounds at frequencies below 4–5 kHz.

Many theories have been advanced to explain coding of sound periodicity in the neural impulse train. Among the earliest theories was Wever's volley theory (1949), which postulated that nerve impulses in primary auditory nerve fibers were *time-locked* to such periodic sounds as, for instance, pure tones. Because the maximal discharge rate of auditory nerve fibers is low compared to the highest discriminable pitch, it was assumed that nerve impulses were not necessarily evoked in synchrony with each sound wave, but rather, depending on the frequency of the sound, with every first wave, second wave, third wave, and so on. These theories were based on experiments in which oscilloscopic registrations of the discharges were superimposed. The oscilloscope sweep was locked to the periodicity of the pure tone used as stimulus. Since then, more sophisticated methods using statistical signal analysis have revealed that phase-locking to a sound wave occurs as a statistical phenomenon by virtue of the probability of firing being greatest at a certain phase of a periodic sound. The frequency of sine waves and the waveform of complex sounds are reflected in the discharge pattern of auditory nerve fibers.

It should be noted that temporal information available in the discharge pattern of single auditory nerve fibers represents the *filtered* and half-wave rectified version of the sound. The filter consists of the transmission characteristics of the outer and the middle ear and especially those of the basilar membrane. Both amplitude and phase characteristics of filtering are essential to the reproduction of the waveform of a complex sound.

A question related to temporal analysis concerns the perception of the fundamental frequency in a periodic sound where the spectral components at the fundamental frequency are missing. Examples of such sounds are complex periodic sounds giving rise to a harmonic spectrum. Even when the fundamental spectral components of such a sound are removed (either by filtering or manipulation of the impulse train), the fundamental frequency is still clearly heard. More than 100 years ago, Seebeck (1843) showed that the pitch of a periodic impulse train could be discerned clearly even though the fundamental frequency was absent and only higher harmonics reached the ear. Such a procedure eliminated the appearance of the fundamental frequency in the traveling wave of the basilar membrane, but temporal periodicty was preserved. It was concluded that pitch discrimination must have its basis in the temporal pattern of the sound since there was no spectral energy present at the fundamental frequency of the sound. It has been confirmed in numerous studies that a distinct pitch sensation exists in accordance with the time envelope of the sound (see Small, 1970). However, it could not be excluded from these earlier experiments that a spectral component equal to the fundamental frequency of the sound was generated within the ear itself through nonlinearities. If this was the case, pitch could be based on the place hypothesis. This possibility was first raised by Helmholtz (1863) who suggested that the fundamental frequency of periodic sound was restored somewhere in the ear by the generation of the difference tones between the harmonics. (The difference between the harmonics of a periodic sound is always equal to the fundamental frequency of the sound.) This particular distortion product has since been studied in experiments using two tones (Chapter 2, p. 155). From such experiments, distortion product appears to be an unlikely basis for spectral pitch discrimination of periodic sounds in the absence of fundamentals because of its strong dependence on the intensity of the primary spectral components.

Subsequent work (Goldstein, 1970) shows that some distortions, particularly the difference between $2f_1$ and f_2, may be sufficiently intense to explain pitch perception on the basis of spectral cues in periodic sounds where no energy exists at the fundamental frequency. In cancellation experiments, certain distinct properties of the $2f_1 - f_2$ component were observed which favor the hypothesis that it might play a role in pitch perception. The level required to cancel the $2f_1 - f_2$ component was only about 15 dB below the level of the fundamentals (f_1 and f_2). The simple difference frequency $f_2 - f_1$, however, increases with intensity and is weak or even nonexistent at low sound levels. The intensity of the $f_1 - f_2$ component, is practically independent of the frequency separation of the two tones, whereas that of the $2f_1 - f_2$ component decreases with the frequency separation of the primary tones.

The possibility of a complex sound's fundamental frequency being restored within the ear was studied in psychoacoustic experiments, particularly those by Schouten (1938, 1940), who found strong evidence against this hypothesis. He found that (a) pitch perception could not be cancelled by a pure tone of a frequency equal to the periodicity of the pulse train; (b) pitch perception could not be masked by noise which might completely mask perception of a sinusoidal tone of the same frequency as that perceived; and (c) that beats could not be detected when a pure tone of a frequency slightly different from that of the periodicity of the pulse train was added. Schouten therefore concluded that the pitch of a periodic pulse train (up to approximately 500 Hz) must be based on the time pattern of the sound rather than on its spectrum. This finding is supported by the fact that the pitch of a low-frequency pure tone—sinusoidally amplitude modulated with a frequency lower than the carrier—is not always related to the spectral components or the envelope of the sound. It is related rather to the fine structure of the temporal pattern of the wave (Schouten, Ritsma, & Cardoza, 1962; Ritsma, 1970). Other studies, which have tried to distinguish between a temporal and a spectral basis for pitch perception, have come to the same conclusion (Licklider, 1962; Small, 1970). Experiments of much greater sophistication leave little doubt that pitch can indeed be discriminated under many circumstances on the basis of the temporal pattern of a sound (Nordmark, 1970).

The physiological basis of this particular temporal pitch must have, as a prerequisite, an interaction between two or more of the spectral components of sound. The width of the spectral filter of the auditory periphery determines the number of harmonics of the sound that will interact. If the filter accommodates more than one partial tone, the periodicity of the fundamental frequency is preserved in the half-wave rectified version of the sound because the difference between each harmonic component of such a sound is equal to the repetition frequency of the sound. In such a case, periodicity or repetition rate of the sound's fundamental frequency will modulate the discharge rate in single auditory nerve fibers as long as fundamental frequency is below 4–5 kHz.

Interactions between individual harmonics of a periodic sound are not necessarily restricted to cases where more than one harmonic is contained within one cochlear spectral filter. Interactions between the time pattern of firings of auditory nerve fibers may occur at secondary neurons of the cochlear nucleus where the nerve fibers from different locations on the basilar membrane converge, if the periodicity of the sound has been preserved in the discharge pattern of single auditory nerve fibers leading thereto.

Place versus Temporal Principle for Pitch Perception

When discussing the place versus the temporal principle as a basis for pitch perception, it is important to address the question of the dynamic range of the ear. The ear can function psychophysically over a range of more than 100 dB. However, the range of pure tone sound intensities for which the discharge rate of single auditory nerve fibers is a function of the sound intensity rarely exceeds 20–30 dB above threshold. Above that level, the discharge rate reaches a saturation level, as shown on p. 145. It is therefore difficult to understand how the discharge rate of a given nerve fiber can convey information about the intensity and frequency distribution of sound to the higher nerve center within a range of 80–100 dB. Not only can information about the intensity of a sound not be conveyed by the discharge rate of single nerve fibers, but also the hypothesis that frequency discrimination occurs according to the place principle seems to be incorrect if it is based on the mean discharge rate in single auditory nerve fibers. It has been shown that neural syncrony to the time pattern or the envelope of complex sounds is preserved in the discharge pattern of single auditory nerve fibers over a much larger range of sound intensities (Sachs & Young, 1979; Javel, 1980, 1981). Findings indicate that it may not be the average discharge rate but the synchronized discharges that convey information to higher neural centers about the sound energy in different frequency bands. Another possibility of conveying information about absolute sound intensity is through the number of nerve fibers that are activated by a sound. There is an increased spread of activity over nerve fibers with increasing intensity. This fact may contribute information about the intensity of simple sounds, such as one or a few pure tones, which have narrow spectral peaks, but it is difficult to understand how information about the intensity of broadband sounds, such as white noise, can be communicated by the discharge rate of single auditory nerve fibers. One may, therefore, speculate that there are a few nerve fibers in which the discharge rate is a function of the sound intensity over a much larger range than those commonly studied. Why these nerve fibers have not been recorded is unknown, but one reason may be that microelectrodes are selective in regard to types of cells or fibers. It may also be that the central nervous system obtains its information about the intensity of a steady sound from the first few milliseconds of the sound and then stores that information until a change in the intensity occurs. That latter hypothesis is supported by the finding that there are cells in the dorsal cochlear nucleus that respond to small spectral peaks and valleys over a large range of intensities of 80–100 dB SPL (Evans, 1977), indicating the existence of some form of automatic gain control operated from neighboring frequencies. It may also be noted that in reproducing the amplitude modulation of a pure tone, nerve cells of the cochlear nucleus have a broad dynamic range, which is further broadened when a second, unmodulated tone of a frequency coinciding with the inhibitory area of the unit is added (Møller, 1975). Thus, intensity coding seems more efficient when the intensity of sounds changes, than when it remains at a constant level.

Single auditory nerve fibers respond to tones and noise stimulation with an average discharge rate that is remarkably constant from 20–30 dB above threshold, and throughout most of the remaining physiological intensity range of the auditory system (Kiang et al., 1965). This evidence speaks against the hypothesis that the mean synchronized discharge rate of individual fibers serves to communicate information about distribution of energy in different frequency bands, a hypothesis which must be correct for the place principle of frequency discrimination to hold true. However, studies on the phase-locked response to pure tones (Rose, Hind, Anderson, & Brugge, 1971) and to noise (Møller, 1977, 1978a,b) reveal that the temporal pattern of the filtered version of sound is preserved in the time pattern of the discharge over a considerably larger range of intensities. Thus, modulation of the discharge rate remains nearly constant over a range from about 20 dB above threshold to 80–90 dB above threshold (Møller, 1977). Studies also indicate that some form of an automatic gain control mechanism may exist, which tends to keep discharge rate at a constant level, regardless of the average amplitude of hair cell stimulation (basilar membrane deflection) (Møller, 1977, 1979). The fact that the rate of phase-locked discharges is practically unaffected by the average stimulus intensity may support the hypothesis that temporal coding occurs in the nervous system.

Two findings support the fact that discharge rate is kept nearly constant through an automatic gain control and not by a saturation mechanism. Histograms of responses of single auditory nerve fibers to low-frequency tones have the shape of (half-wave rectified) sinusoids (Rose et al., 1971) over a large range of sound intensities. In addition, responses to noise (Møller, 1977, 1978a, b) show little distortion as judged from the high values of coherence spectra over a large range of sound intensities obtained in these studies. It is likely that automatic gain control mechanism operates with a certain time constant and therefore does not distort the waveform of test sound waves (Rose et al., 1971).

In contrast to other hypotheses which proposed a more or less linear relationship between neural excitation (discharge rate) and sound intensity, it now seems more likely that there exists an active regulation mechanism which tends to keep the discharge rate within narrow limits while maintaining the phase-locking of the neural discharges to the waveform of the sound. If this mechanism takes a certain time to regulate the sensitivity, it would facilitate transmission of changes in a sound since it keeps the neurons operating, with regard to discharge rate, at an optimal level and independent of the absolute level of the sound. Such an automatic gain control mechanism would also facilitate the temporal analysis of sound with regard to furnishing higher centers with information about frequency through periodicity coding, but it makes it difficult to conceive how spectral analysis of steady-state sounds could exist as explained in the place principle.

Studies by Sachs and Leng (1980) using synthetic vowel sounds indicate the importance of the temporal pattern of such sounds in conveying information, not only about the fundamental (vocal cord) frequency, but also about the frequencies of the peaks in the spectrum (formant frequencies).

There are other aspects of intensity coding in single auditory nerve fibers that merit attention. In Chapter 2 it was mentioned that stimulus response curves of single auditory nerve fibers obtained using single pure tones showed a lower saturation level at stimulus frequencies above the fiber's best frequency (Sachs & Abbas, 1974). The same phenomenon occurs in cells of the cochlear nucleus (Møller, 1969) and the superior olivary nucleus (Guinan, Guinan, & Norris, 1972). However, its importance in coding the distribution of excitation along the basilar membrane is not known.

By examining the response areas of single auditory nerve fibers, as they appear in the shape of frequency threshold curves, it is clear that low-frequency sounds at physiological levels (55–80 dB above threshold) not only activate the fibers that are tuned to the frequencies of the sounds but also activate the low frequency tails of the response areas of fibers that are tuned to higher frequencies. That means that a large number of high-frequency tuned nerve fibers will respond to low-frequency sounds. It has been shown that responses of such auditory nerve fibers with high characteristic frequencies to low-frequency sounds are phase-locked to the periodicity of sound (Kiang, Sachs, & Peake, 1967). That means that not only nerve fibers with low characteristic frequencies below 4–5 kHz can transmit periodicity information about sounds at physiological levels to higher centers, but that essentially all nerve fibers have that property.

If our understanding of spectral analysis is that it can identify where along the basilar membrane a sound produces the maximal deflection, such analysis is unlikely to account for frequency discrimination below 4–5 kHz because (a) maximal deflection shifts location as a function of sound intensity; and (b) the range within which the discharge rate of individual auditory nerve fibers is a function of sound intensity (deflection of the basilar membrane) is extremely limited. It appears that at frequencies above 4–5 kHz, frequency must be communicated in terms of information about which nerve fiber(s) are carrying the highest discharge rate. The place principle thus seems the only available explanation for frequency discrimination above the frequency where the time pattern of a sound no longer can be recovered from the discharge patterns of single auditory nerve fibers.

In psychoacoustic experiments, uncertainty will always exist as to whether a particular judgment about pitch has been arrived at, based on the time pattern or on the spectrum. Although neurophysiological experiments may be less subject to this difficulty, they cannot directly provide information about whether a particular sound is perceived on the basis of temporal or spectral coding. If the time pattern of a sound cannot be recovered in the discharge patterns of individual auditory nerve fibers, it is unlikely that temporal cues have played an important role in frequency discrimination. Moreover, that the temporal pattern of the neural discharge frequency is modulated with the periodicity of a sound does not in itself prove that temporal cues are important in that regard.

The inability to elucidate the process by which the central nervous system ultimately decodes information about pitch has led to a certain bias against a temporal basis for pitch perception (Whitfield, 1970). This, however, is a matter of accessibility. The well-established tonotopic organization of the nuclei of the ascending auditory pathway, up to and including the auditory cortex, has been seen as providing the foundation for the decoding of data concerning pitch on a spectral basis (place principle). Nerve fibers or cells carrying the highest level of excitation of the basilar membrane would, it is conjectured, be identified by the higher auditory centers from their anatomical locations. Although attractive and seemingly reasonable, this remains, for the present, mere conjecture. Our understanding of the interpretative functions of the central nervous system as a whole is unfortunately limited. We can favor neither spectral nor temporal cues on the basis of any presumed capacity of the central nervous system to decode them.

If pitch perception is based on temporal information, however, the waveform of a sound must somehow be recoded in the neural discharge pattern at a relatively early point in the ascending auditory pathway; otherwise, the temporal pattern would become blurred in passing through the various neural relay stations en route to the cerebral cortex. The best evidence that such coding does take place in the auditory system is provided by studies of binaural hearing.

It has been discovered that units in the superior olive and the inferior colliculus are responsive to the time interval between clicks applied to the left and right ears. These units are also sensitive to the phase difference between two pure tones applied, one to each ear, at frequencies up to about 1500 Hz.

Since psychoacoustic experiments have clearly demonstrated that certain factors in directional hearing are based on the time relation between the arrival of sound at the two ears, the auditory nervous system is obviously capable of detecting small time intervals. In comparing pitch perception with directional hearing, it should not be overlooked that, in the latter case, we are measuring the time between two events at different locations (i.e., the two ears), whereas in pitch perception the task is to determine the time lapse between events occurring at the same location. The significance of this is not yet known, but the results of some psychoacoustic experiments indicate that the mechanism responsible for both the detection of the binaural time difference and for the periodicity of monaural sounds may be similar to that responsible for pitch discrimination in monaural listening (Nordmark, 1970). There is, however, a limit above which temporal coding of frequency becomes unlikely. Because binaural phase effects are no longer perceived in the frequency range above 1500 Hz, the upper limit of periodicity coding has often been estimated at about that frequency.

Under certain circumstances, the transient units in the cochlear nucleus described on p. 133 may perform a similar task with regard to time intervals between successive events in a monaural sound. Results of studies previously described (p. 202) using noise as stimuli and a cross-correlation technique also point toward a dominance of temporal coding of frequency in the discharge patterns of individual auditory fibers.

An upper limit for periodicity detection at frequencies of about 4–5 kHz is obviously in effect. Although we are not certain that the upper limit of periodicity coding is that high in man, it is worth noting that the frequency range throughout which this coding occurs in the auditory nerves of many animals does, in fact, include most of the audible range at which pitch perception is of any practical importance in man. The implication of the fact that the range in question constitutes but a small part of the audible frequency range of the experimental animals used is not clear.

What Information Does the Ear Convey to the Central Auditory Nervous System in the Case of Steady-State Sounds?

It may be of interest to consider what information the inner ears gives to the central nervous system in the cases of a few elementary stimuli and to discuss how this information may be transformed by a simple neural mechanism known to exist in the various brain nuclei of the ascending auditory pathway—namely, spatial integration, which occurs when many nerve fibers converge into a single neuron.

In the case of a steady, *single pure tone* with a frequency below about 5 kHz, it is assumed that it gives rise to a phase-locked discharge pattern in single auditory nerve fibers. In addition, the discharge rate will be greater in certain primary auditory fibers than in others, in accordance with the assumed greater excitation of hair cells in the region where the traveling wave has its highest amplitude. With regard to the processing of the time pattern of the neural discharges evoked by such a sound, it may be assumed that many nerve fibers carry the same periodicity. Spatial integration no doubt will enhance this periodicity. The integrated activity of many primary fibers can thus be expected to convey to higher centers rather precise information about the periodicity of the sound.

Click stimuli, another frequently used sound in both physiological and psychoacoustic experiments, gives in many ways a completely different pattern of neural discharges in primary nerve fibers. Such a sound may be assumed to excite all, or at least many, nerve fibers to an almost equal degree—those with high characteristic frequencies being excited first. As the traveling wave progresses along the cochlear partition, nerve fibers with lower characteristic frequencies are excited as well. As was described earlier in this chapter, the temporal structure of the discharge pattern of each nerve fiber is phase-locked to the damped oscillation of the basilar membrane. The frequency of this damped oscillation is different for each fiber and equal to the characteristic frequency of each fiber. Spatial integration of the activity of many fibers thus largely extinguishes the periodicity of this damped oscillation, leaving only the initial transient. The temporal pattern of the firing in that case provides information to higher centers about the characteristic frequency of individual fibers, and thus yields information about the identity of the fibers activated rather than information about the property of the stimulus. In the case of stimulation with repetitive clicks, the periodicity of the click presentation is enhanced by spatial averaging, particularly since the firing of fibers with high characteristic frequency is relatively well synchronized to the click sounds. A population of fibers with low characteristic frequency (corresponding to nerve fibers terminating at apically located hair cells) will fire less synchronously than those that terminate at basal hair cells because of the slower propagation of the traveling wave in the apical part of the cochlea.

Random noise (continuous) has the same average spectral distribution as brief clicks, and therefore such a sound will also excite all nerve fibers to nearly the same degree. Individual hair cells are assumed to be stimulated with a bandpass-filtered version of the noise—the bandpass filter being the spectral selectivity of the basilar membrane. The time pattern of the discharges of single nerve fibers in that case is also phase-locked to the characteristic frequency of the fibers. Since the output of a bandpass filter, in response to noise, is similar to a pure tone (sinusoid) that is amplitude modulated with a random signal, it may be assumed that the discharges of single auditory nerve fibers are phase-locked to the frequency of that sinusoid. The frequency of that sinusoid is equal to the characteristic frequency of the individual fiber and the degree of modulation is a function of the bandwidth of the filter. Upon stimulation with transient sounds as well as broadband random noise, the discharge pattern of individual single auditory nerve fibers may therefore be assumed to be modulated with a periodicity corresponding to the characteristic frequency of the fiber. The temporal code can thus only reflect which fibers are activated (i.e., the spectral extension of the sound). When the activities of many fibers with different characteristic frequencies are integrated, the fundamental periodicity is lost and only the random signal that modulates the excitation of the individual hair cells remains.

Phase-locking of the discharges of single auditory fibers to the low-frequency periodicity of a sound may thus supply the central nervous system with two types of information, depending on the stimulus: (a) the periodicity of the sound (if it is periodic); and (b) the identity of the fibers that are being activated by a transient stimulation or stimulation with broadband random noise.

It may also be appropriate to mention that these hypotheses and experimental results regard continuous sounds. Natural sounds are usually time-varying, with energy distributed over various parts of the audible frequency range, and they differ in many ways from the sounds used in our experiments. Little is known about the processing of such sounds in the ear and in the auditory nervous system, but the results obtained so far indicate that the transformations are extensive and cannot generally be predicted on the basis of knowledge about the response to continuous simple sounds such as pure tones and click sounds.

Conclusion

Discrimination of complex sounds cannot be explained on the basis of spectral analysis alone nor on the basis of temporal analysis of the sound alone. Although it is experimentally well documented that the ear does some type of spectral analysis, it is not known what role spectral analysis plays in the perception of complex sounds. The spectral analysis that occurs in the ear has often been referred to as a Fourier analysis but the analysis that takes place in the ear is rather different from the mathematical operation in Fourier analysis that concerns separation of a complex wave in a number of sine waves with different amplitudes and phases. Analysis performed in the ear can be better described as a separation of information contained in different frequency bands. This information may then be processed in different channels of the ascending auditory system. This processing may consist of determination of the total energy in a certain band and analysis of the time structure of the signal carried in each of these channels. The spectral analysis that occurs in the inner ear establishes the frequency width of these channels, and it seems as if both the width and the frequency location of these bands are functions of the sound intensity. These channels should not be regarded as a few discrete channels where many nerve elements carry information within a certain band of frequencies, but instead should be conceptualized as a continuum, along which the center frequency of each nerve element varies in very small steps. There is extensive interaction between nerve elements with different center frequencies throughout the ascending auditory pathway.

As we have seen in this chapter, temporal analysis may be more important for processing of complex sounds than the simple determination of energy in different frequency bands—at least for low frequencies. The time pattern of the waveform as it is coded in the discharge pattern may play an important role in the perception of natural sounds. For high frequencies (above 4–5 kHz), it would seem that the only important clue might be the energy in a certain frequency band, but it may be the time pattern of the envelope of the signals carried in these different channels that is perceptually important.

This and other principles of processing of complex sounds in the ascending auditory pathway will be treated in the following chapter.

References

Békésy, G. von Über die Schwingenen der Schneckentrennwand biem Präparat und Ohrenmodell. Akustiche Zeitschrift, 1942, 2(7), 173–186. [The vibration of the cochlear partition in anatomical preparations and in models of the inner ear.] Journal of the Acoustical Society of America, 1949, 21, 233–245.

Békésy, G. von Über die resonanz Kurve und die Abklingzeit der verschiedenen Stellen der Schneckentrennwand. Akustiche Zeitschrift, 1943, 8, 66–76. [On the resonance curve and the decay period at various points on the cochlear partition.] Journal of the Acoustical Society of America, 1949, 21, 245–254.

Békésy, G. von Microphonics produced by touching the cochlear partition with a vibrating electrode. Journal of the Acoustical Society of America, 1951, 23, 29–35.

Békésy, G. von Description of some mechanical properties of the organ of corti. *Journal of the Acoustical Society of America*, 1953, 25, 770. a

Békésy, G. von Shearing microphonics produced by vibrations near the inner and outer hair cells. *Journal of the Acoustical Society of America*, 1953, 25, 786. b.

Billone, M. C. *Mechanical stimulation of cochlear hair cells*. Unpublished master's thesis, Northwestern University, Evanston, Ill., 1972.

Boer, E. de. Correlation studies applied to the frequency resolution of the cochlea. *Journal of Auditory Research*, 1967, 7, 209–217.

Boer, E. de. Reverse correlation. I. A heuristic introduction to the technique of triggered correlation with applications to the analysis of compound systems. *Proceedings of the Koninklijke Nederlandse Akademie van Wetenschappen*, 1968, 71, 472–486.

Boer, E. de. Reverse correlation. II. Initiation of nerve impulses in the inner ear. *Proceedings of the Koninklijke Nederlandse Akademie van Wetenschappen*, 1969, 129–151.

Campbell, F. W., Cleland, B. G., Cooper, G. F., & Enroth-Cugell, C. The spatial selectivity of visual cells of the cat. *Journal of Physiology* (London), 1969, 203, 223–235.

Dallos, P. *The auditory periphery*. New York: Academic Press, 1973. a

Dallos, P. Cochlear potentials and cochlear mechanics. In A. R. Møller (Ed.), *Basic mechanisms in hearing*. New York: Academic Press, 1973. b

Dallos, P., Billone, M. C., Durrant, J. D., Wang, C. Y., & Raynor, S. Cochlear inner and outer hair cells: functional differences. *Science*, 1972, 177, 356–358.

Dallos, P., & Cheatham, M. A. Compound action potentials (AP) tuning curves. *Journal of the Acoustical Society of America*, 1976, 59, 591–597.

Dallos, P., & Harris, D. Properties of auditory nerve responses in absence of outer hair cells. *Journal of Neurophysiology*, 1978, 41, 365–383.

Duifhuis, H. Cochlear nonlinearity and second filter: Possible mechanism and implications. *Journal of the Acoustical Society of America*, 1976, 59, 408–423.

Evans, E. F. The effects of hypoxia on the tuning of single cochlear nerve fibers. *Journal of Physiology* (London), 1974, 238, 65–67.

Evans, E. F. Cochlear nerve and cochlear nucleus. K. D. Keidel & W. D. Neff (Eds.), *Handbook of sensory physiology*. New York: Springer-Verlag, 1975. a

Evans, E. F. Normal and abnormal functioning of the cochlear nerve. Sound reception in mammals. *Symposium of the Zoological Society of London*, 1975, 37, 133–165. b

Evans, E. F. Frequency selectivity at high signal levels of single units in the cochlear nerve and nucleus. In E. F. Evans & J. P. Wilson (Eds.), *Psychophysics and physiology of hearing*. New York: Academic Press, 1977.

Evans, E. F., & Wilson, J. P. Frequency sharpening of the cochlea: The effective bandwidth of cochlear nerve fibers. *Proceedings of the 7th International Conference on Acoustics*, 1971, 3, 453–460.

Evans, E. F., & Wilson, J. P. Frequency selectivity of the cochlea. In A. R. Møller (Ed.), *Basic mechanisms in hearing*. New York: Academic Press, 1973.

Flanagan, J. L. Models of approximating basilar membrane and displacement. Part II: Effects of middle ear transmission and some relations between subjective and physiological behavior. *Bell Systems Technical Journal*, 1962, 41, 959–1009.

Flock, A. Contractible proteins in hair cells. *Hearing Research*, 1980, 2, 411–412.

Flock, A., & Cheng, H. C. Actin filaments in sensory hairs of inner ear receptor cells. *Journal of Cell Biology*, 1977, 75, 339–343.

Gilbert, A. G., & Pickles, J. O. Responses of auditory nerve fibers in the guinea pig to noise bands of different width. *Hearing Research*, 1980, 2, 327–333.

Goblick, T. J., & Pfeiffer, R. R. A test for cochlear linearity from cochlear nerve spike discharges in response to combination click stimuli. *Journal of the Acoustical Society of America*. 1968, *44*, 363.

Goldstein, J. L. Aural combination tones. In R. Plomp & G. F. Smoorenburg (Eds.), *Frequency analysis and periodicity detection in hearing*. Leiden: A. W. Sijthoff, 1970.

Greenwood, D. D., & Goldberg, J. M. Response of neurons in the cochlear nucleus to variations in noise bandwidth and to tone-noise combinations. *Journal of the Acoustical Society of America*, 1970, *47*, 1022–1040.

Guinan, J. J., Guinan, S. S., & Norris, B. E. Single auditory units in the superior olive complex. I. Responses to sounds and classification based on physiological properties. *International Journal of Neuroscience*, 1972 *4*, 101–120.

Hall, T. L. Cochlear models: Evidence in support of mechanical nonlinearity and second filters (A review). *Hearing Research*, 1980, *2*, 455–464.

Helmholtz, H. L. F. von. [*Die Lehre .von den Tonempfindungen als physiologische Grundlage für die Theorie der Musik*] (A. J. Ellis, Ed. and Trans.). New York: Dover, 1954. (Originally published, 1863.)

Houtgast, T. Psychophysical experiments on "tuning curves" and "two-tone inhibition," *Acustica*, 1973, *29*, 168–179.

Hudspeth, A. J., & Corey, D. P. Sensitivity, polarity and conductance change in response of vertebrate hair cells to controlled mechanical stimuli. *Proceedings of the National Academy of Sciences USA*, 1977, *74*, 2407–2411.

Huggins, W. H. A phase principle for complex-frequency analysis and its implications in auditory theory. *Journal of the Acoustical Society of America*, 1952, *24*, 582–589.

Javel, E. Coding of AM tones in the chinchilla auditory nerve: Implications for the pitch of complex tones. *Journal of the Acoustical Society of America*, 1980, *68*, 133–146.

Javel, E. Suppression of auditory nerve responses I: Temporal analysis, intensity effects and suppression contours. *Journal of the Acoustical Society of America*, 1981, *69*, 1735–1745.

Johnstone, B. M., & Boyle, A. J. F. Basilar membrane vibration examined with the Mössbauer technique. *Science*, 1967, *158*, 389–390.

Khanna, S. M., Sears, R. E., & Tonndorf, J. Some properties of longitudinal shear waves: A study by computer simulation. *Journal of the Acoustical Society of America*, 1968, *43*, 1077–1084.

Kiang, N.-Y. S., Watanabe, T., Thomas, E. C., & Clark, L. F. *Discharge patterns of single fibers in the cat's auditory nerve*. Cambridge, Massachusetts: MIT Press, 1965.

Kiang, N.-Y. S., Sachs, M. B., & Peake, W. T. Shapes of tuning curves of single auditory nerve fibers. *Journal of the Acoustical Society of America*, 1967, *42*, 1341–1342.

Licklider, J. C. R. Periodicity and related auditory process models. *International Audiology*, 1962, *1*, 11–36.

Lim, D. Fine morphology of the tectorial membrane. *Archives of Otolaryngology*, 1972, *96*, 199–215.

Lim, D. Cochlear anatomy related to cochlear micromechanics: A review. *Journal of the Acoustical Society of America*, 1980, *67*, 1686–1695.

Manley, G. A. Cochlear frequency sharpening—a new synthesis. *Acta Otolaryngologica* (Stockholm), 1978, *85*, 167–176.

Møller, A. R. Unit responses in the cochlear nucleus of the rat to pure tones. *Acta Physiologica Scandinavica*, 1969, *75*, 530–571.

Møller, A. R. Unit responses in the cochlear nucleus of the rat to noise and tones. *Acta Physiologica Scandinavica*, 1970, *78*, 289–298. a

Møller, A. R. Studies of the damped oscillatory response of the auditory frequency analyzer. *Acta Physiologica Scandinavica*, 1970, *78*, 299–314. b

Møller, A. R. Dynamic properties of excitation and inhibition in the cochlear nucleus. *Acta Physiologica Scandinavica*, 1975, *93*, 442–454.

Møller, A. R. Frequency selectivity of single auditory nerve fibers in response to broadband noise. *Journal of the Acoustical Society of America*, 1977, *62*, 135–162.

Møller, A. R. Responses of auditory nerve fibers to noise stimuli show cochlear nonlinearities. *Acta Otolaryngologica* (Stockholm), 1978, *86*, 1–8. a

Møller, A. R. Frequency selectivity of the peripheral auditory analyzer studied using broadband noise. *Acta Physiologica Scandinavica*, 1978, *104*, 24–32. b

Møller, A. R., & Nilsson, H. G. Inner ear impulse response and basilar membrane modelling. *Acustica*, 1979, *41*, 258–262.

Müller, J. Handbuch der Physiologie des Menchen, II. Abt. 1, Pp. 393–483, 1837. [Elements of physiology.] (W. Baly, Trans.) Philadelphia: Lea & Blanchard, 1843.

Nilsson, H. G. A comparison of models for sharpening of the frequency tuning in the cochlea. *Biological Cybernetics*, 1977, *28*, 177–181.

Nordmark, J. Time and frequency analysis. In J. V. Tobias (Ed.), *Foundations of modern auditory theory* (Vol. 1). New York: Academic Press, 1970.

Peake, W. T., & Ling, A. Basilar membrane motion in the alligator lizard: Its relation to tonotopic organization and frequency selectivity. *Journal of the Acoustical Society of America*, 1980, *67*, 1736–1745.

Pfeiffer, R. R., & Kim, D. O. Considerations of nonlinear response properties of single cochlear nerve fibers. In A. R. Møller, (Ed.), *Basic mechanisms of hearing*. Academic Press, New York: 1973.

Pfeiffer, R. R., & Kim, D. O. Cochlear nerve fiber responses: Distribution along the cochlear partition. *Journal of the Acoustical Society of America*, 1975, *58*, 867–869.

Pfeiffer, R. R., & Molnar, C. E. Cochlear nerve fiber discharge patterns: Relationship to cochlear microphonics. *Science*, 1970, *167*, 1614–1616.

Pierson, M., & Møller, A. R. Some dualistic properties of the cochlear microphonics. *Hearing Research*, 1980, *2*, 135–150. a

Pierson, M., & Møller, A. R. Effect of modulation of basilar membrane position on cochlear microphonics. *Hearing Research*, 1980, *2*, 151–162. b

Rhode, W. S. Observations of the vibration of the basilar membrane in squirrel monkeys using the Mössbauer technique. *Journal of the Acoustical Society of America*, 1971, *49*, 1218–1231.

Rhode, W. S. Some observations on cochlear mechanics. *Journal of the Acoustical Society of America*, 1978, *64*, 158–176.

Rhode, W. S., & Robles, L. Evidence from Mössbauer experiments for nonlinear vibration in the cochlea. *Journal of the Acoustical Society of America*, 1974, *54*, 588–596.

Robles, L., Rhode, W. S., & Geisler, C. D. Transient response of the basilar membrane measured in squirrel monkeys using Mössbauer effect. *Journal of the Acoustical Society of America*, 1976, *59*, 926–939.

Rose, J. E., Hind, J. E., Anderson, D. J., & Brugge, J. F. Some effects of stimulus intensity on responses of auditory nerve fibers in the squirrel monkey. *Journal of Neurophysiology*, 1971, *34*, 685–699.

Russell, I. J., & Sellick, P. M. The tuning properties of cochlear haircells. In E. F. Evans & J. P. Wilson (Eds.), *Psychophysics and physiology of hearing.* New York: Academic Press, 1977.

Russell, I. J., & Sellick, P. M. Intracellular studies of hair cells in the mammalian cochlea. *Journal of Physiology* (London), 1978, *284*, 261–290.

Sachs, M. B., & Abbas, P. J. Rate versus level functions of auditory nerve fibers in cats: Tone burst stimuli. *Journal of the Acoustical Society of America*, 1974, *56*, 1835–1847.

Sachs, M. B., & Young, E. D. Encoding of steady-state vowels in the auditory nerve: Representation in terms of discharge rate. *Journal of the Acoustical Society of America*, 1979, *66*, 470–479.

Sachs, M. B., & Young, E. D. Effects of nonlinearities on speech encoding in the auditory nerve. *Journal of the Acoustical Society of America.* 1980, *68*, 858–873.

Schouten, J. F. The perception of subjective tones. *Proceedings of the Koninklijke Nederlandse Akademie van Wetenschappen*, 1938, *41*, 1086–1093.

Schouten, J. F. The residue and the mechanism of hearing. *Proceedings of the Koninklijke Nederlandse Akademie van Wetenschappen*, 1940, *43*, 991–999.

Schouten, J. F., Ritsma, R. J., & Cardoza, B. L. Pitch of the residue. *Journal of the Acoustical Society of America*, 1962, *34*, 1418–1424.

Seebeck, A. Üeber die Sirene. *Annalen der Physik und Chemie*, 1843, *60*, 449–481.

Sellick, P. M., & Russell, I. J. Two tone suppression in cochlear hair cells. *Hearing Research*, 1979, *1*, 227–238.

Sellick, P. M., & Russell, I. J. The response of inner hair cells to basilar membrane velocity during low frequency auditory stimulation in the guinea pig cochlea. *Hearing Research*, 1980, *2*, 439–445.

Small, A. M. Pure tone masking. *Journal of the Acoustical Society of America*, 1959, *31*, 1619–1625.

Small, A. M. Periodicity pitch. In J. V. Tobias (Ed.), *Foundations of modern auditory theory* (Vol. 1). New York: Academic Press, 1970.

Vogten, L. L. M. Pure tone masking. A new result from a new method. In E. Zwicker & E. Terhardt (Eds.), *Facts and models in hearing.* Berlin: Springer-Verlag, 1974.

Wever, E. G. *Theory of hearing.* New York: Wiley, 1949.

Whitfield, I. C. Central nervous processing in relation to spatio-temporal discrimination of auditory patterns. In R. Plomp & G. G. Smoorenburg (Eds.), *Frequency analysis and periodicity detection in hearing.* Leiden: A. W. Sijthoff, 1970.

Wilson, J. P., & Evans, E. F. Grating acuity of the ear: Psychophysical and neurophysiological measures of frequency resolving power. *Proceedings of the 7th International Congress on Acoustics*, 1971, *3*, 397–400.

Wilson, J. P., & Johnstone, J. R. Basilar membrane and middle ear vibration in guinea pig measured by capacitive probe. *Journal of the Acoustical Society of America*, 1975, *57*, 705–723.

Zwicker, E. Die Verdeckung von Schmalbandgerauschen durch Sinustone. *Akuts. Beih. (Acustica)*, 1954. *1*, 415–420.

Zwicker, E. On the psychoacoustical equivalent of tuning curves. In E. Zwicker & P. Terhardt (Eds.), *Facts and models in hearing.* Berlin: Springer-Verlag, 1974.

Zwicker, E., & Schorn, K. Psychoacoustical tuning curves in audiology. *Audiology*, 1978, *17*, 120–140.

Zwislocki, J. J. A possible neuromechanical sound analysis. *Acustica*, 1974, *31*, 354–359.

Zwislocki, J. J. Symposium on cochlear mechanics: Where do we stand after 50 years of research? *Journal of the Acoustical Society of America*, 1980, *67*, 1679–1685.

Zwislocki, J. J., & Kletsky, E. J. Tectorial membrane: A possible effect on frequency analysis in the cochlea. *Science*, 1978, *204*, 639–641.

Zwislocki, J. J., & Kletsky, E. J. Micromechanics in the theory of cochlear mechanics. *Hearing Research*, 1980, *2*, 505–512.

CHAPTER 4

Coding of Complex Sounds in the Auditory Nervous System

Introduction

In Chapters 2 and 3, coding of relatively simple, steady-state sounds in the discharge pattern of single auditory nerve fibers and cells in the ascending auditory nervous system were considered. A fundamental question now arises: To what extent can we extrapolate results obtained using simple sounds to provide information about how the auditory system handles and analyzes complex sounds, particularly time-varying sounds? This subject was briefly discussed in Chapter 2. Studies have revealed that it is generally not possible to predict how the auditory system processes time-varying complex sounds on the basis of knowledge about processing of simple and steady-state sounds. It is thus not possible to predict the response to complex sounds of single nerve fibers and single nerve cells in the auditory system on the basis of results obtained using pure tones (with constant amplitude and frequency) or click sounds. However, relatively few neurophysiological studies have been conducted using broadband, complex sounds as stimuli.

This matter will be considered in some detail in this chapter as well as results obtained from single cell recordings in the nuclei of the ascending auditory pathway in animal experiments. Stimuli in these experiments were sounds that changed either in frequency or amplitude or, more recently, speechlike sounds. Most of the results originate from studying the cochlear nucleus and the auditory nerve.

RELATIONSHIP BETWEEN STIMULUS AND PERCEPTION OF SOUND

Figure 4.1 shows the basic components of a sensory system and their function. The stimulus is conducted to a receptor. In the auditory system, the conductive system is the ear canal, middle ear, and cochlea. Sounds that do not reach the receptor, because they are outside the frequency range of the conductive apparatus, naturally cannot be perceived. The receptor is limited with regard to sensitivity. Sounds that are below a certain intensity do not evoke any response. The frequency selectivity of the cochlea results in the separation of a complex sound into frequency bands, each activating a different group of sensory cells.

The ascending auditory nervous system (like other sensory systems) transforms the information and enhances certain features, whereas other features are suppressed. Signal processing becomes more complex as one moves toward the primary auditory cortex where there undoubtedly also occurs an extensive processing of information that reaches the cortex from the medial geniculate body. In addition, there is an extensive feedback system by which information that reaches a certain level of the

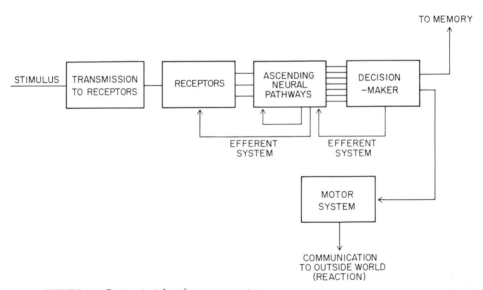

FIGURE 4.1. Basic principles of a sensory system.

ascending auditory pathway is likely to control the signal processing in a more peripheral nucleus or even at the level of the sensory receptor.

These features that are extracted from the sound by the auditory nervous system, about which we know very little, form the basis for an animal's reaction to a sound. The simplest reaction to a sound may be flight, which occurs when the sound is interpreted as indicating an approaching threat to the animal. The more complex reactions to sounds that man may experience span the whole range from simple muscle reactions (startle response) to comprehension of spoken words with possibilities of recalling the message from memory.

What we can study most in the nervous system is the discharge pattern of single nerve cells. We can compare, for example, the wave pattern of a stimulus with the discharge pattern of auditory nerve fibers in order to study the transformation that occurs in the inner ear and the hair cells' transduction of the vibration of the basilar membrane. By comparing the discharge pattern of fibers in the auditory nerve with that of cells in the cochlear nucleus, one can study the transformation of the information that occurs in the cochlear nucleus. Similar studies may provide information about the transformation that occurs in other of the different relay nuclei of the ascending auditory pathway.

The auditory system, like other sensory systems, is capable of extracting various features of a sound and discriminating the difference between many stimuli. Probably, only a few of these features are used by any particular animal. Which of these properties is used by a certain animal depends on the *revlevance* of certain sounds to that particular animal (Granit, 1978).

It is evident from many psychoacoustic studies that the auditory system can extract such features of simple sounds as frequency and intensity, and transform them into perceptive qualities called pitch and loudness. However, it is not clear if the neural processing that underlies perception, recognition, and comprehension of complex sounds that have relevance to man is the same neural processing that underlies the discrimination of simple sounds. Although both types of sounds are processed by the same auditory system, it is neither obvious nor likely that the basis and principles for recognition of different sounds occurring at high brain levels are the same for complex sounds as they are for simple sounds such as pure tones and click sounds that have no meaning. These two types of sounds may even be subjected to different types of processing in the ascending auditory pathway.

The fact that we can discriminate small changes in the frequency of a tone or train of impulses shows that the nervous system has the ability to extract information about small changes in the frequency of a tone or train of pulses. It is, however, not known if the same neural processing is used as a basis for discrimination of natural sounds such as speech. Results of experiments on scaling of pitch and loudness provide us with a subjective scale (measure) of loudness or pitch of a pure tone, but it does not imply that the same neural processing which is the basis for loudness and pitch discrimination is used in discrimination of speech sounds. That means that pitch, loudness, and other sensations we may have in response to simple sounds may not reflect the type of analysis that takes place in the auditory system in discrimination of relevant sounds, namely, speech.

In other words: It is not directly obvious that recognition of speech is based on the same type of neural processing as is the perception of the pitch of a simple sound presented in isolation in a situation where the test subject is asked a specific question about its pitch.

It must be emphasized that we know very little about how neurophysiological results are related to perception. Thus, we do not know which properties of the discharge pattern are essential for perception. To a great extent, this is a result of our almost total lack of information about how sensory information is decoded in the nervous system. Consequently, we cannot differentiate the functional importance of various types of coding of sounds in the discharge pattern of single nerve cells.

In summary, we know relatively well what the ear and the auditory system can discriminate with regard to simple sounds, but we know very little about how the auditory system analyzes complex time-varying sound. We know next to nothing about how complex sounds are discriminated. It is reasonable to ask then why we have devoted most of our efforts to studying the auditory system using such simple sounds.

VALIDITY OF ELECTROPHYSIOLOGICAL STUDIES

A great deal of our knowledge about the function of the auditory nervous system, as well as of the nervous system in general, is based on interpretations of recordings of the electrical events in the brains of experimental animals, usually under anesthesia. Recordings have been made of the electrical activity in a large population of nerve cells using large electrodes (evoked responses) or electrical events in single nerve cells (unit responses). Recordings from single nerve cells are made using microelectrodes and it is the discharge pattern that is usually studied. Recordings from single nerve cells are regarded as easier to interpret and are usually assigned a greater value in describing the function of the part of the nervous system under study than are gross responses. One difficulty in interpreting results obtained from recordings of single nerve cells stems from the fact that, due to technical difficulties, it is not possible to record from more than one (or in rare cases, a few, chosen by random) nerve cells at a time. In contrast, the present and generally accepted concept of the function of the nervous system is that it is the combined, integrated activity of many neurons that is significant.

Normally the discharge pattern of a single cell is more or less random. In neurophysiological experiments, repeated presentation of the same stimulus is used in connection with some form of averaging of the response to obtain a stable estimate of the response to a certain stimulus. The nervous system most likely integrates the activity of many neurons, thereby decreasing statistical variations in the response of individual nerve cells so that a single stimulus presentation gives rise to a stable response. We have some confidence that averaging the response of a single nerve element to many stimulus presentations may give a relevant picture of the processing of various sounds in the nervous system. However, it is far from certain that such is the case. It is important to bear this in mind when results obtained using recordings from single nerve cells are interpreted.

Also, it should be kept in mind that the discharge pattern that can be recorded from a single nerve cell or nerve fiber is only one of many electrical phenomena in nerve cells, and that activation of a nerve cell gives rise to a series of chemical processes. We do not know what role these chemical processes and other electrical phenomena play in the communication between nerve cells and, subsequently, in the signal processing in the nervous system.

Another significant obstacle in studies of the nervous system in experimental animals using electrophysiological methods is the influence of anesthesia, usually assumed to impair synaptic transmission. When recording from single nerve cells in the ascending auditory nervous system, two types of influence may be discerned. One is the direct influence on the ascending (afferent) pathway and the other is the influence on the descending (efferent) pathway. The descending system is assumed to have the potential of modulating or modifying the information transfer in the ascending pathway. When information transfer in the descending pathway is influenced by anesthesia, for example, impulse activity in the ascending pathway may be affected in a number of ways. Since we have practically no understanding of the function of the descending system and its interaction with the ascending pathway, it is not possible to estimate the effect of impairment of transfer of information in the descending system on impulse activity in the ascending pathway.

Naturally the solution to this problem is to use unanesthetized animals. However, the technique is complicated and the quantitative yield of responses in such situations is low. In addition, there are other difficulties with the techniques, one of the greatest being the influences of the animal's attention and its degree of alertness on the response pattern. These two variables are especially difficult to control or monitor.

In the following sections we will discuss the results of recording from single nerve cells in various parts of the nuclei of the ascending auditory pathway and from single fibers of the auditory nerve in response to various types of complex sounds.

Responses to Broadband Sounds in the Auditory Periphery

In Chapter 2 it was briefly shown that the intensity-frequency area within which a nerve cell in the cochlear nucleus responded when stimulated with tones with varying frequency depended strongly on the rate with which the frequency of the tone varied. In Chapter 3, it was shown that frequency resolution of the periphery of the auditory system, as it appears in the discharge pattern of single auditory nerve fibers, is different for different sound intensities. In these experiments, the sound was a broadband noise. Generally these results showed that the responses to complex sounds cannot be deduced from results obtained using simpler sounds, such as pure tones. These and other discrepancies between results obtained using different types of sounds will be considered in detail in the present chapter.

The discharge pattern of single auditory neurons to complex sounds has been studied systematically only in recent years. Existing knowledge is limited and data related to the higher brain nuclei and complex sounds, such as speech, are scanty.

Most investigations, which have been made into the responses of time-varying sounds, have been restricted to relatively simple sounds, such as amplitude and frequency-modulated tones and noise. Only the cochlear nucleus, which is the most peripheral part of the auditory nervous system, has been studied systematically. A few studies have been published on the responses of neurons in the superior olive and inferior colliculus. There have been only brief reports on responses of nuclei of the medial

geniculate and the auditory cortex to similar stimuli. Generally, these
studies show that responses to complex sounds undergo drastic transformations in the auditory nervous system. The response pattern even to
simple sounds, such as amplitude and frequency-modulated sounds, cannot be predicted on the basis of knowledge about the response to simpler
sounds, such as tone bursts and click sounds.

INFLUENCE OF BANDWIDTH

The response of single auditory nerve fibers and cells in the cochlear
nucleus to steady-state bandpass-filtered noise is a function not only of
the intensity of the noise but also of its bandwidth. Up to a certain bandwidth the discharge rate increases with increasing bandwidth of the noise
(Evans & Wilson, 1973; Greenwood & Goldberg, 1970; Møller, 1970).
This is a result of the spectral integration performed by frequency-selective elements in the periphery of the auditory system. If the system had
behaved like a linear filter and if the discharge rate had been a function
only of the energy of the noise passed through the filter, the discharge rate
would have increased with increasing bandwidth up to a certain bandwidth value. Above this bandwidth value, the discharge rate would be
constant, independent of the bandwidth. The bandwidth of the noise
stimulus where the discharge rate would be expected to reach a plateau is
equal to the width of the tuning curves of the cells (Greenwood &
Maruyama, 1965; Greenwood & Goldberg, 1970).

Many cells in the cochlear nucleus behave in a different manner
(Greenwood & Goldberg, 1970). Whereas the discharge rate usually increases as the bandwidth of a noise stimulus is increased up to the width of
the cell's frequency threshold curve (FTC), the discharge rate does not
maintain a plateau value for noise bandwidth above the width of the FTC,
as would be expected if the system had functioned as a linear filter. On the
contrary, the discharge rate usually decreases as the bandwidth of the
noise is increased beyond the width of the cell's FTC. In fact, many nerve
cells in the cochlear nucleus do not respond at all to a steady broadband
noise. This is illustrated in Figure 4.2, where the discharge rate of typical
cochlear nucleus cells is shown as a function of the bandwidth (Greenwood & Goldberg, 1970). The left graph shows the relationship of the
spike count of a single cell in the cochlear nucleus to the intensity of a tone
at CF and a narrowband noise centered at CF. The right-hand graph
shows the variation of the spike count with the bandwidth of a bandpass-
filtered noise centered at the unit's CF, maintaining a constant spectrum
level.

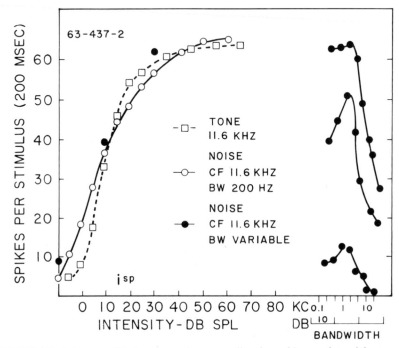

FIGURE 4.2. *Left curves:* Discharge rate of a nerve cell in the cochlear nucleus of the cat in response to tones and noise of different bandwidths. *Right curves:* Discharge rate in response to bandpass-filtered noise as a function of bandwidth for three different intensities. (From Greenwood & Goldberg, 1970.)

Although there is a large degree of variation between units in the response to noise, it is consistently found that the effectiveness of noise as an excitatory stimulus decreases with its bandwidth above a certain bandwidth value. The higher the sound level of the stimulus is, the narrower the bandwidth value is at which the decrease in discharge rate will begin to occur.

In some units the shape of the PST histograms of responses to bursts of noise also changes as a function of bandwidth. Figure 4.3 shows PST histograms of responses to bursts of noise of a different bandwidth value (Greenwood & Goldberg, 1970). The figure shows that the shape of the PST histogram varies as a function of bandwidth. This is an indication that the adaption to noise sounds is related to the bandwidth of the sounds. It may be concluded from these studies that steady broadband noise is not an efficient stimulus for most cells in the cochlear nucleus.

Broadband noise is also a less efficient stimulus in auditory nerve fibers

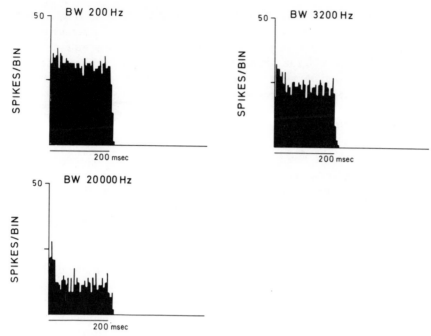

FIGURE 4.3. Poststimulus time histograms of the response of a nerve cell in the cochlear nucleus to noise of different bandwidth. (From Greenwood & Goldberg, 1970.)

than are pure tones and narrowband noise. Thus several investigators (Ruggero, 1973; Gilbert & Pickles, 1980) have shown that the discharge rate of primary nerve fibers increases with increasing bandwidth—keeping the spectral density constant up to a certain value, above which it decreased—of the noise. The mean bandwidth value at which maximal response was obtained was only 34% of the fiber's CF (Gilbert & Pickles, 1980). The reason for the decreased response to broadband noise was assumed to be two-tone inhibition. As the bandwidth of the noise was increased, its spectrum covered more of the inhibitory areas surrounding the fiber's excitatory response area.

RESPONSE TO DIFFERENT TYPES OF BANDPASS-FILTERED SOUNDS

The average discharge rate of neurons in the cochlear nucleus is not just a function of the frequency or spectrum and the intensity of a sound, but it is also related to the time structure of a sound. Not only will the time pattern of the neural discharge of these neurons depend on the time structure of the sound, but two sounds with equal mean energy and an equal spectrum envelope may not evoke the same average (mean) discharge rate. An example of such a difference is seen when the discharge rate of single neurons in the cochlear nucleus in response to bandpass-filtered noise is compared with that evoked by bandpass-filtered click sounds. Figure 4.4 shows discharge rate as a function of stimulus intensity for a typical cell in the cochlear nucleus of a rat. Noise (right graph) and click sounds (left graph) with the same bandwidth are shown (Møller, 1972). The results in this figure were obtained by passing wideband noise and wideband clicks through the same bandpass filter. The different curves in each graph represent different bandwidths. As the bandwidth is increased, the stimulus response curves for noises of different bandwidth are parallel and shift toward the larger response. Whereas that can be explained by the spectral integration of the ear—which results in sound being a more efficient stimulus when its bandwidth is increased—the responses to bandpass filtered clicks are more complex and cannot be explained by spectral integration. For sound levels that evoke less than one discharge per filtered click, the stimulus response curves are parallel and shifted as a function of bandwidth in a fashion similar for noise. For higher sound levels, the stimulus response curves are not parallel as they were for bandpass-filtered noise. Narrowband transients may evoke more nerve impulses than broadband stimuli. This means that increasing the bandwidth of filtered click sounds above a certain value in fact results in a decrease in discharge rate and what appears to be a decrease in the efficiency of the stimulus in evoking a neural response. The reason for this paradoxical response to bandpass-filtered click sounds becomes obvious when the time pattern of the sound is considered and when it is understood that one of the properties of a neuron is its refractoriness. During its absolute refractory period a neuron cannot fire, regardless of how strong the stimulus is. Its threshold is increased immediately after the end of this absolute refractory period, after which the threshold gradually approaches its normal value. Wideband clicks have a shorter duration than narrowband clicks.[1] Therefore, a wideband click may be over before the end of the re-

[1] When an impulse such as a rectangular wave is passed through a bandpass filter, the result is a damped oscillation where the frequency of the oscillation is equal to the center frequency of the bandpass filter. The exact shape of the envelope of the oscillation depends on which type of bandpass filter is used but, in general, the duration of the oscillation is a function of the bandwidth of the filter. The broader the filter (the larger the bandwidth), the shorter the damped oscillation of the filter will be. It is presumed that the rectangular wave has a duration that is shorter than the time for one-half period of the damped oscillation of the bandpass filter.

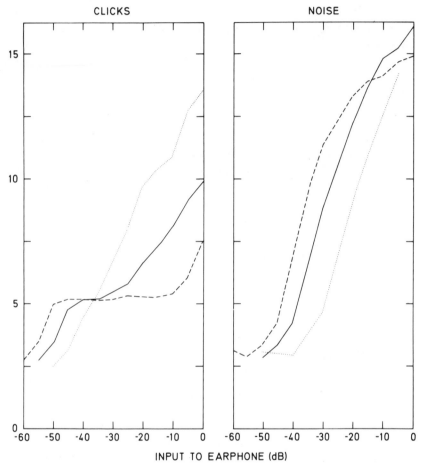

CLICKS

NOISE

INPUT TO EARPHONE (dB)

FIGURE 4.4. Response of a single nerve cell in the cochlear nucleus of a rat to bandpass-filtered clicks (left graph) and bandpass-filtered white noise (right graph). The sounds were presented in bursts of 50-msec duration and the average number of discharges evoked per sound burst is given as a function of sound intensity. The filtered clicks were generated by filtering 30-μsec-long rectangular pulses with a repetition rate of 100 ppsec (5 clicks in each 50-msec burst). Both clicks and noise were passed through the same bandpass filter thus making the spectrum envelope of both signals identical. The center frequency of the filter was 5.2 kHz, equal to the CF of the unit. The different curves represent different bandwidths of the filter. Dotted line: 250 Hz; solid line: 750 Hz; and dashes: 1900 Hz. (From Møller, 1972.)

fractory period of the neuron and may never evoke a second nerve impulse independent of how high its initial energy is. Narrowband filtered clicks that have a duration longer than the refractory period, on the other hand, may evoke more than one nerve impulse if the energy is sufficient (Figure 4.4).

Responses to Tones and Noise with Rapidly Varying Frequency (Sweep Tones and Sweep Noise)

In neurophysiological experiments, the frequency selectivity of the auditory periphery has traditionally been assessed using pure tones. In fact, most available results are based on the threshold of single fibers or nerve cells to stimulation with pure tones. We have learned a great deal about the function of the auditory periphery from these results. In fact, results presented in the form of frequency threshold curves (FTC) have become more or less established measures of the functional frequency selectivity of the auditory periphery. However, the selectivity near threshold for a pure tone may not provide information about the functional selectivity to natural sounds (i.e., sounds at a level well above threshold and with a more or less broad spectrum). In addition, natural sounds usually vary their spectra rapidly.

COCHLEAR NUCLEUS

As described briefly in Chapter 2 (p. 171); the response of single neurons in the cochlear nucleus to tones of slowly varying frequency is radically different from their response to tones with rapidly varying frequency. These units responded to tones with rapidly varying frequency within a narrower frequency range than they did to tones with slowly varying frequency. This resulted in the clustering of more nerve impulses within a narrower frequency range of the tone with rapidly varying frequency compared to that with a slowly varying frequency. Thus, increasing the rate of change in frequency of a tone resulted in a redistribution of the discharges. More nerve impulses were evoked within a narrower frequency range than when the frequency of the tones was varied slowly.

Systematic studies of responses to tones, the frequencies of which varied at different rates, have revealed that most cells in the cochlear nucleus show a similar change in the response pattern to tones as the rate of change in frequency is varied. However, the rate of frequency

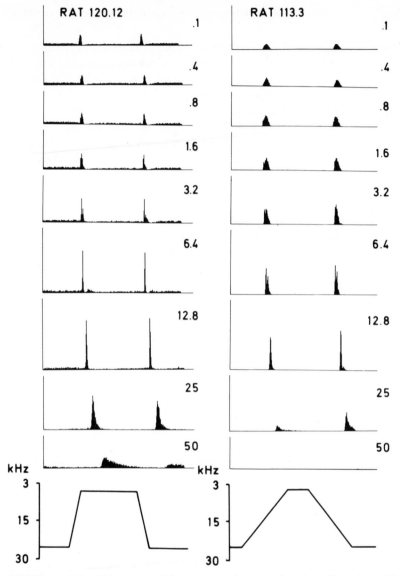

FIGURE 4.5. Period histograms of the response of two single nerve cells in the cochlear nucleus of the rat to tones the frequencies of which were varied at different rates and in accordance with the scheme seen below each of the two columns. The sweep rate is indicated by numbers on each histogram (in sweeps per second). The sound intensity at the unit's CF was 45 dB SPL (CF 22 kHz, left graph) and 50 dB SPL (CF 15 kHz, right graph), which corresponds to approximately 25 dB and 15 dB above threshold, respectively. (From Møller, 1972.)

change—at which the greatest enhancement occurred—varied from cell to cell from 400 kHz/sec to 15 MHz/sec, as did the degree of increase in the height of the histograms of the response to tones with time-varying frequency. The rate of change in frequency that produced the largest peaks in the histograms does not seem to be related to the CF of the neuron (Møller, 1969, 1971, 1972). The height of the peaks corresponding to a falling tone frequency in many cells reached a higher value than that corresponding to increasing frequency. Figure 4.5 shows histograms of responses to trapezoidally frequency-modulated tones of two units in the cochlear nucleus of a rat. The legend numbers give the repetition rates of the modulation that was varied from .1 to 50 Hz. The corresponding rate of change in frequency is 30 kHz/sec to 15 MHz/sec for the neuron depicted in the left graph and 3 kHz/sec to 1.5 MHz/sec for the neuron depicted in the righthand graph. The right peak in each histogram represents the response to decreasing frequency and the left peak represents the response to increasing frequency.

Since the frequency of the tones of varying frequency stay within the frequency range of the neurons—as predicted by the frequency tuning curve of the cell—for a period of time that is a direct function of the rate of change in frequency, there is a possibility that the change in the pattern of response to tones with rapidly varying frequency is a result of adaptation. Results of experiments in which the frequency of the tones was kept constant and equal to the CF of the nerve cell while the sound was interrupted repetitively so that the tone was on for a constant fraction of the repetition period do not show a similar sharpening of the peaks in the histograms as shown in response to tones with rapidly changing frequency (Figure 4.6D). Stimulation with chopped tones of constant fre-

---

FIGURE 4.6. Relative height of the peaks in the period histograms of the response to sweep tones (upper graphs, FM, A, B, C) and average discharge rate (lower graphs) shown as a function of sweep rate and rate of change in frequency of the tone. The different graphs are for different sound pressure levels (indicated by legend numbers). Dashed lines represent down-going frequency sweeps, and solid lines, up-going sweeps. Graph D shows the relative height of the peaks in the histograms of the response to chopped tones of a constant frequency (equal to the unit's CF) as a function of repetition rate (upper graph). The different symbols indicate intensity of the tonebursts: open circles, 65 dB; triangles, 45 dB; and closed circles, 25 dB above threshold at the cell's CF. The lower graph shows the corresponding spike counts contained in the peaks in the histograms of the response to amplitude-modulated sounds. The symbols correspond to those in the upper graph. (From Møller, 1971.)

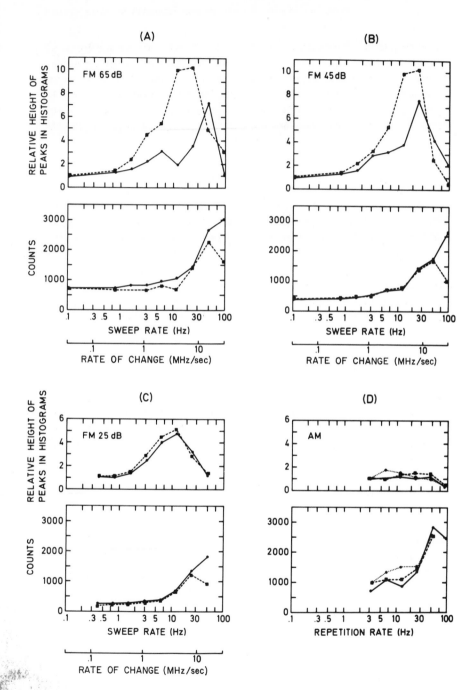

265

quency produced histograms similar to those obtained using tones with slowly varying frequency when the tones were relatively long. However, when the repetition rate of the tones with constant frequency was increased to mimic the situation with tones of rapidly varying frequency, the histograms never showed the same increase in peak height as was the case for the response to tones with rapidly varying frequency. It may therefore be assumed that the response pattern—increase in height of the peaks in the histograms—is specifically caused by the change in frequency and not a result of adaptation.

This is supported by the finding that the height of the histograms of the response to tones with rapidly varying frequency does not depend significantly on the repetition rate with which the frequency is changed, as is seen in experiments where the rate of change in frequency is varied independently of the repetition rate. The results of such experiments show that it is essentially the rate of change in frequency of a tone that gives rise to the change in the height of the histograms of the response to sweep tones and not the fact that the tone "stays" within the response area of the unit for a shorter period when the frequency of the tone is varied rapidly. The individual graphs in Figure 4.6 show the response to sweep tones of different intensities. It should be noted that the increase in peak height is greatest for tones whose intensities are well above the unit's threshold. This "enhancement" of the response does not saturate at a stimulus level below that of natural sounds, as does the average discharge rate.

Another characteristic feature of responses to sweep tones is dependency on the direction of the frequency sweep. A down-going frequency sweep (decreasing frequency) always gives rise to a larger increase than a rising frequency sweep does in the height of the peaks in the histograms. In addition, maximal value of the peaks in the histogram usually occurs at a somewhat lower rate of change in the frequency of the stimulus tone for down-going sweeps than for up-going sweeps (increasing frequency).

Since natural sounds usually appear in a background of some type of noise, it is interesting to consider how the response pattern changes when a background noise is added. When a background noise becomes strong enough, it reduces the increase in height of the peaks in the histograms that occurs with an increase in rate of change in frequency of the stimulus tone—but even in the case where a strong background noise is presented together with the sweep tones, there is a clear increase in peak height with increasing rate of change in frequency (Møller, 1974c).

The conclusion that may be drawn from these experiments is that the response to sweep tones is consistently different from what would be expected on the basis of the responses to steady-state tones or tones with slowly varying frequency. Whereas results obtained using tones with slowly varying frequency are in accordance with those obtained using tones of constant frequency presented in bursts, results obtained using tones with rapidly varying frequency could not have been predicted on the basis of the unit's response to tones with constant or slowly varying frequency. It might be inferred that these nerve cells in the cochlear nucleus show a greater degree of frequency selectivity in response to tones with rapidly varying frequency than they do to tones with constant or slowly varying frequency. However, it should be noted that these data do not provide any direct information about frequency discrimination or the ability to respond to each of several spectral components individually.

Different groups of neurons in the cochlear nucleus may be discerned on the basis of their responses to tones of rapidly varying frequency. The response pattern to sweep tones just described and shown in Figures 4.5 and 4.6 are typical for one common group of neurons. Only the value of the rate of change in frequency at which the histograms had their maximal height varies from unit to unit (Møller 1972, 1974c). A few neurons in different groups are characterized by having no apparent frequency selectivity in response to tones with slowly varying frequency selectivity. In spite of this, these neurons had distinct frequency selectivity when stimulated with sound of rapidly varying frequency (Møller 1971, 1972, 1974c). Histograms of the responses of such neurons to tones with slowly varying frequency are nearly flat, whereas distinct peaks develop when the rate of change in frequency is raised above a certain value. Typical histograms of a unit that belongs to this class are shown in Figure 4.7. The frequency of the tone was varied at different rates in a manner similar to that used to obtain the results shown in Figure 4.5, where the frequency of the triangular frequency-modulated tones is shown in the lower insert in the graph.

Another class of neurons in the cochlear nucleus are characterized by showing inhibition within a certain frequency area in response to tones with constant or slowly varying frequency. These units usually have a relatively high spontaneous discharge rate and no excitatory area when stimulated by tones of constant frequency. However, when stimulated with tones of rapidly varying frequency, these units show an excitatory response as the frequency of the tone is varied at a certain rate (Figure 4.8). An increase in firing rate may then be seen within a certain range of frequencies of the stimulating tone. Despite the fact that these units only display inhibition in response to tones with slowly varying frequency

UNIT 85.5   CF 3.5 KHz

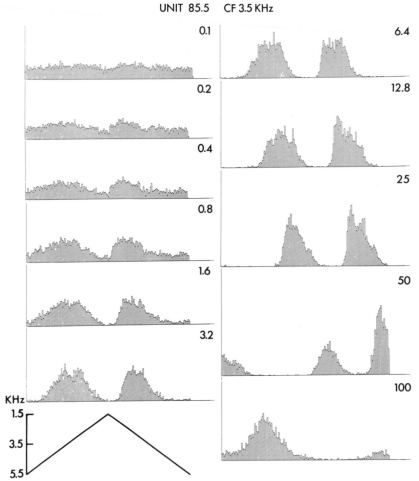

FIGURE 4.7. Period histograms of the responses to sweep tones of a unit in the cochlear nucleus of the rat that did not show any appreciable frequency selectivity for constant tones or tones with slowly varying frequency. (From Møller, 1969.)

they thus respond to tones with quickly changing frequencies in a manner very similar to that described for the more common units which show frequency selectivity also to tones of steady frequency.

These results show that the responses of single neurons in the cochlear nucleus to tones of rapidly varying frequency cannot be predicted on the basis of knowledge of the response to tones of constant frequency. These

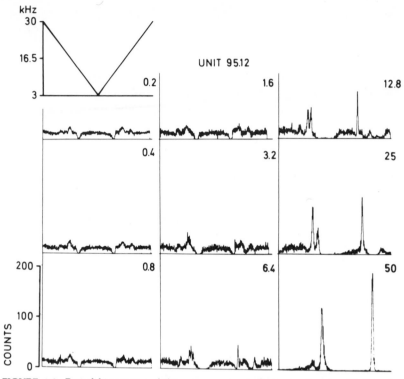

FIGURE 4.8. Period histograms of the responses to sweep tones of a nerve cell in the cochlear nucleus. The responses to tones of constant or slowly varying frequency were mostly inhibitory. (From Møller, 1974c.)

results also indicate that neurons in the cochlear nucleus show specificity with regard to parameters other than the steady-state frequency of a tone, namely, to change in frequency.

   Although the above results were obtained using sounds more complex than those normally used to determine the frequency selectivity of neural responses in the auditory system, these sweep tones represent simple sounds when compared with natural sounds. An example of sounds more complex than tones, yet well defined and not as complex as speech sounds, is a band of noise, the center frequency of which is varied at different rates. When the center frequency is varied at different rates, the response varies in a manner similar to that described for sweep tones. That means that a redistribution of firing, similar to that described for sweep tones, occurs also when noise bands, the center frequencies of which are varied rapidly, are used as stimuli. Thus histograms of the responses to such sounds show an increase in height as the rate of change of the center frequency of the noise is increased up to a certain rate of frequency change, above to which the height decreases again. There is one difference, however, between the response to sweep tones and the response to noise bands of varying frequency. The average number of discharges evoked by noise bands usually also increases when the rate of change in the center frequency is increased.

Histograms of typical responses from a nerve cell in the cochlear nucleus to bandpass filtered noise, the center frequency of which is varied up and down at different rates, are shown in Figure 4.9 together with histograms of the responses to sweep tones for the same nerve cell. The different columns represent noises of different bandwidth. The figure shows an increase in the height of the peaks in the histograms similar to that shown for sweep tones. Although the enhancement of the peaks of the responses to noise is somewhat smaller than that response to tones, the maximal peak height occurs at about the same rate of change in center frequency of the noise. Even nerve cells that do not respond well (some do not respond at all) to stimulation with bandpass filtered noise with steady center frequency when the bandwidth of the noise exceeds a certain value, respond vigorously to bandpass filtered noise when the center frequency is varied rapidly (see Chapter 2). Nerve cells that respond poorly or not at all to bandpass filtered noise of a certain bandwidth thus will respond to the bandpass filtered noise if its center frequency is varied at a certain rate. The response pattern, under such circumstances, gives a somewhat different impression than the response to sweep tones. The response to noise bands with slowly varying center frequency is very small or nonexistent. However, above a certain sweep rate, the response pattern is similar to that of units that also respond to noise with slowly varying center frequency. Examples of responses from a nerve cell that did not respond to broadband noise when the center frequency was slowly varied are shown in Figure 4.10. The unit had a distinct response and even a distinct frequency selectivity when the center frequency of the noise was varied rapidly. This figure also shows the histograms of the responses to stimulation with tones and noise with narrower bandwidth, from which the response to steady-state sounds and to sounds with slowly varying frequency may be determined. The frequency was varied at two different rates: .1 and 6.4 Hz, corresponding to 30 kHz/sec and 1.92 MHz/sec,

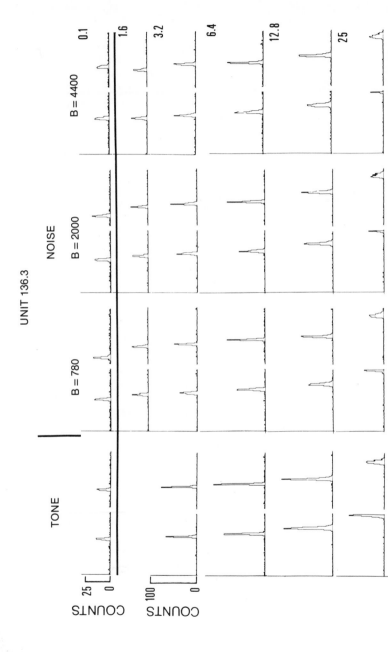

FIGURE 4.9. Period histograms of the responses of a single nerve cell in the cochlear nucleus to stimulation with sweep tones (left column) and bands of noise the center frequency of which was varied in a fashion similar to that in which the sweep tones were varied. The different columns are responses to noises of different bandwidth (780, 2000, and 4400 Hz). The histograms were expanded in a way similar to those seen in Figure 2.40, to show only 16% of each total sweep cycle. The CF of the unit was 17.6 kHz and the intensity of the tone was 54 dB SPL (25 dB above threshold). (From Møller, 1974c.)

271

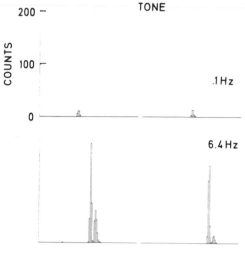

TONE

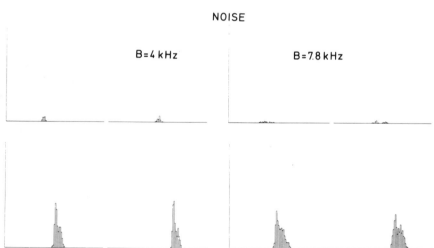

NOISE

FIGURE 4.10. Period histograms of the response to sweep tones and sweep noise of a cell in the cochlear nucleus that showed very little response to broadband noise with constant or slowly varying center frequency. The CF of the unit was 13.2 kHz and the intensity of the tone was 55 dB SPL corresponding to 20 dB above threshold. Only 16% of the total sweep is shown in each histogram. (From Møller, unpublished.)

respectively. The histograms are displayed in a manner similar to that in which the histograms are displayed in Figure 2.40 and only the area around the peaks is shown. In the unit for which responses are shown in Figure 4.10, a 7.8 kHz wideband noise does not give rise to any significant response when the center frequency is varied slowly (.1 Hz). However, when the center frequency is varied rapidly (6.4 Hz) the unit responds vigorously, even showing a pronounced frequency selectivity. Pure tones and noise with a narrower bandwidth also respond at slow sweep rates.

Because the broadband noise gives such a small response at slow sweep rates, the difference in height of the peaks in the histograms for high and low sweep rates becomes much greater for the broadband noise than for tones and noise with a narrower bandwidth. When the relative heights of the histograms of the responses to bandpass-filtered noise of varying center frequency are plotted as a function of sweep rate as in Figure 4.6, the response to broadband noise appears to be greater than that to pure tones (Figure 4.11). This is because these units respond poorly to noise with slowly varying center frequency. (It is interesting to note that at a level as peripheral as the cochlear nucleus, specificity can be discerned in the discharge pattern with regard to parameters such as rate of change in frequency.)

Thus, there is no doubt that nerve cells in the cochlear nucleus generally have a preference for sounds of changing frequency. It is evident that different classes of nerve cells respond more or less specifically to changes in frequency. Thus, there seems to be a separation of sounds according to their different characteristics, and different sounds are transmitted in different neural channels to a higher center. In the examples just mentioned, such separation seems to be based on the rate of change in frequency of the sound. It should be noted that most natural sounds have spectra that change more or less rapidly in time. The fact that this sensitivity to change in frequency persists over a large range of sound intensities, being more pronounced at physiological sound levels than at or just above threshold is important as is the fact that background noise does not significantly alter the response. These findings indicate that this property just described may be of physiological importance.

PRIMARY AUDITORY NERVE FIBERS

Information on the coding of rapid changes in frequency in single auditory nerve fibers is limited, and only a few studies have been published. Available results show that responses to rapid changes in frequency in the firing pattern of single auditory nerve fibers are enchanced little, in contrast to the enhancement that occurs in the cochlear nucleus. When histograms of the responses to sweep tones are displayed in a manner similar to that shown in Figure 4.5 for the cochlear nucleus, only a

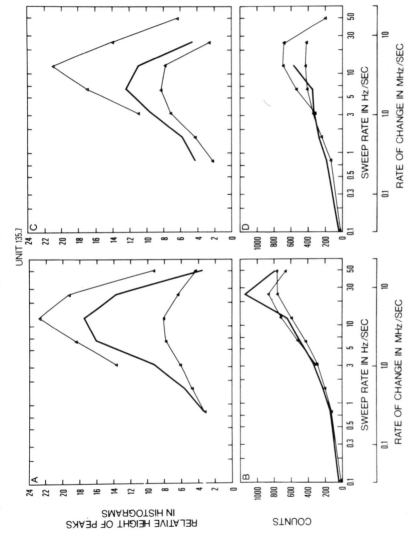

FIGURE 4.11. Height of the peaks in the period histograms of the response to tones, the frequencies of which were varied at different rates, compared to the response to bands of noise, the center frequencies of which were varied at different rates. Solid lines without symbols: Tone; thin lines: Noise. Bandwidth 2000 Hz, lower curves, 7800 Hz, upper curves. Lower graphs show discharge rate. *Left graphs:* Decreasing frequency; *right graphs:* increasing frequency. (From Møller, 1974c.)

slight increase in peak height is shown with increasing rate of change in frequency of the stimulus tone (Sinex & Geisler, 1981).

FUNCTIONAL SIGNIFICANCE OF THE
RESPONSE TO FAST CHANGES IN FREQUENCY

Of the common natural animal sounds, including those made by man, it is obvious that both the songs of the birds and human speech are typical sounds containing rapid changes in frequency. Sounds emitted by the flying bat and used in its echolocation may be the clearest example of a frequency-modulated natural sound. Sounds produced by bats vary according to species, but typically they are short bursts of a high frequency —nearly pure—tone, the frequency of which falls about one octave in a few millisceonds (typically 80–40 kHz in less than 5 msec).

To a great extent, bats rely on their echolocalization system both for navigation in flight and for searching and catching prey. Thus, hearing is critical for survival. There are other animals, such as the dolphin and the oilbird, that use similar systems, but with completely different sounds.

Since the echolocation sounds of the flying bat characteristically contain rapid changes in frequency, considerable attention has been devoted to the neural processing of sounds (tones) with rapidly varying frequency in these animals. Suga and his colleagues (Suga, O'Neill, & Manabe, 1979a and O'Neill & Suga, 1979b) found that the mustache bat, which they studied intensively has specific areas in the auditory cortex that respond to certain features of the echo of the sound emitted. It is believed that the neurons in these areas function to detect important features of the sound (feature detectors).

Thus, there are neurons that respond specifically to sound with a certain delay (which is directly related to the target range). These neurons are sensitive to the spectral composition of the sound in such a way that they can identify the echo of the sound emitted by the bat itself, ignoring the sound originating from fellow bats. Suga and his colleagues found that these neurons respond only if the sound contains the proper set of harmonics.

The fact that responses to tones with rapidly varying center frequencies evoke discharges from the cochlear nucleus units within a narrower range of time may have influenced the character of the sounds produced by the flying bats for their echolocalization. Although not all bats produce only frequency-modulated tones, it is generally thought that in ranging it is the frequency-modulated part of the sound that is important. If so, it is important that the animal is able to determine with great precision the length of time that elapses from evocation of the sound to reaching the ear after having been reflected from an object. Since the response to a frequency-modulated tone has a more distinct representation in time in the discharge pattern of cochlear nucleus units than a tone of constant frequency does, it is possible that the bat's auditory system can determine the time interval between two frequency-modulated sounds more precisely than it can determine the interval between sounds (tones) of constant frequency. In the case of the flying bat this means that frequency-modulated sounds can offer greater precision in ranging only if the rate of change in frequency is within certain values. That may be why bats use them for echolocalization. (There are also bats that are called CF bats because they use continuous tones. For these bats, the doppler effect is likely to be utilized in detecting velocity of targets.)

Coding of Sound with Varying Intensity

In the cochlear nucleus there are many nerve cells that show little to moderate adaptation, but there are also nerve cells that show a high degree of adaptation. In the inferior colliculus only a few nerve cells show a tonic response to tonebursts. Neurons in more centrally located nuclei, such as the medial geniculate and the neurons of the auditory cortex, generally respond only to the onset or offset of a toneburst.

The responses from single auditory nerve fibers and cells in the cochlear nucleus to sinusoidally amplitude-modulated tones were described in Chapter 2. We will now examine the coding of amplitude-modulated tones and noise in the cochlear nucleus in more detail.

AMPLITUDE-MODULATED SOUNDS IN THE COCHLEAR NUCLEUS

The experiments referred to in Chapter 2 all refer to modulation of a single tone or noise. A sine wave contains only one spectral component in contrast to a natural sound which contains many spectral components. When sounds containing more than one spectral component are modulated, the firing pattern of single nerve cells in the cochlear nucleus shows some interesting features that could not be predicted on the basis of the results obtained using single amplitude-modulated tones. Most cells in the cochlear nucleus have both excitatory and inhibitory response areas, as discussed in Chapter 2. If two tones are presented, one which is located within the inhibitory area and one which is located within the excitatory area, amplitude modulation of either of the two will modulate the discharge pattern. However, modulation of the inhibitory tone will result in a modulation of the period histograms of the response that is shifted 180° in phase compared to the results obtained when the excitatory tone is modulated (see Chapter 2).

The results just described were obtained using pure tones modulated with a sinusoidal signal and were in the form of period histograms of the recorded discharge pattern. A more general description of the response to amplitude-modulated sounds may be obtained when broadband noise is used to modulate tones or noise in conjunction with statistical signal analyses of the recorded data. Thus, the responses to tones or noise that are amplitude modulated by a continous broadband noise yield information about the dynamic properties of the system under test.

An approximation of a systems impulse response function may be obtained by using random noise as input to the system and cross-correlating the output of the system with the noise used as input. Temporal integration of the impulse response function yields an estimate of the system's response to a step change in the amplitude of a stimulus (step response function) and the Fourier transform of the system's impulse response function is an estimaste of the system's transfer function. When noise is used to modulate the stimulus, the cross-correlation between the output and the noise used to modulate the stimulus is an estimate of the response to a small, brief hypothetical increase or decrease in the amplitude of the carrier sound. (The output of the system is regarded to be the discharge density.) When the modulation depth is small, as it usually is in these experiments, the impulse response, step response, or modulation transfer function represent the linear portion of the system's response to changes in amplitude. The Fourier transform of the impulse response function becomes an estimate of the system's modulation transfer function. For practical reasons, it is often advantageous to use *pseudorandom noise* instead of ordinary Gaussian noise.

Pseudorandom noise is noise that repeats itself periodically, but that otherwise is similar to random noise. Pseudorandom noise is generated digitally and has properties that give it certain advantages over ordinary random noise. In obtaining the results which will be discussed, pseudorandom noise was used to modulate the amplitude of tones and noise. Because pseudorandom noise repeats itself, it is possible to make a period histogram of the discharge pattern locked to the periodicity of the noise. Such a histogram represents the averaged response to one period of noise and when responses to a sufficient number of periods of the pseudorandom noise are averaged, the modulation of the resulting period histogram is an estimate of the variation in the probability of firing at various instants over one period of the noise. Cross-correlating such period histograms with one period of the noise then yields information about the dynamic properties of the system under test, which is similar to that obtained by cross-correlating the discharge pattern with continuous random noise. A more detailed description of the use of pseudorandom noise in studies of biological systems may be found in Marmarelis and Marmarelis (1978), and specific description of the use of pseudorandom noise in studies of the auditory system may be found in Møller (1973, 1974b).

Using broadband signals, instead of sinusoids to modulate tones and noise, offers advantages other than just being able to obtain a complete modulation transfer function in one experimental trial. Thus, the response pattern to noise-modulated sounds may be studied over a large range of sound intensities using noise-modulated sound without the effect of the discharge pattern being time-locked to the periodicity of the modulation. Using pseudorandom noise to modulate the amplitudes of tones also allows the study of many combinations of carrier tones. Results obtained are either in the form of impulse or step response functions or modulation transfer functions.

Pseudorandom noise has been used to modulate the amplitude of one or two tones in systematic studies of the interaction between excitatory and inhibitory tones in the response of single nerve cells in the cochlear nucleus of the rat. These studies have focused on how small changes in the amplitude of an inhibitory or an excitatory tone are coded in the discharge pattern of these neurons. Figure 4.12 shows typical impulse response and step response functions obtained using tones that were amplitude modulated with pseudorandom noise. The curve marked EXC was obtained when the excitatory tone—frequency equal to the CF of the unit—was modulated. The curve marked INH represents responses obtained when inhibitory tone was modulated and the excitatory tone unmodulated.

The curves in Figure 4.12 show that these functions are mirror images of one another when either of the two tones is modulated. It should be noted that the latency is the same for inhibitory and excitatory responses. This excludes the fact that the inhibition shown in these neurons is not mediated through any additional neurons (interneurons). The results also indicate that the inhibition in these neurons of the cochlear nucleus may have its origin in the cochlea and may be the same as the two-tone suppression in the auditory nerve. Another possibility is that these neurons have both excitatory and inhibitory synapses receiving input from similar auditory nerve fibers. Although this response pattern seems to be the common one of cochlear nucleus units, there are also neurons in the cochlear nucleus with a more complex interplay between excitation and inhibition. However, these have not been studied in sufficient detail.

When the shapes of the impulse and step responses to amplitude-modulated sounds, of the units that have the same latency of inhibition and excitation, are examined in more detail, two different types of neurons may be discerned in the cochlear nucleus. In one type (Figure 4.12, Type I) the shape of the step response shows a gradual monotone decay as is typical for a simple type of adaptation. This indicates that a step increase in sound intensity results in an initial increase in probability of firing followed by a gradual and monotone decrease in probability of firing, similar to that seen in many other parts of the nervous system. In the other type of neurons (Figure 4.13, Type II), the step response is distinctly different. Instead of a monotone decrease, it shows a damped oscillation after the initial increase in firing probability. The amplitude of the damped oscillation exhibited by the Type II neuron, increases with stimulus intensity. It may be insignificant or absent at low stimulus intensities. Responses of the Type I neurons maintain an approximately similar shape over a large range of stimulus intensities (Møller, 1976a).

Whereas the modulation transfer function of the Type I neurons does not change much with increasing sound intensities, that of the Type II neurons changes drastically from being approximately of a lowpass type at low sound intensities to a bandpass type at higher sound intensities. Similar results emerge from experiments using sinusoidally modulated tone stimuli (Møller, 1974a). When stimulated with sinusoidally amplitude-modulated tones, the responses of Type I neurons are almost independent of the modulation frequency up to a certain modulation frequency, above which the response decreases. For Type II neurons, the amplitude of the modulation of the histograms increases with an increase in modulation frequency up to a certain modulation frequency above which it decreases.

The characteristics of the first type of neurons reveal that they adapt in a similar way that other sensory neurons adapt. However, the damped

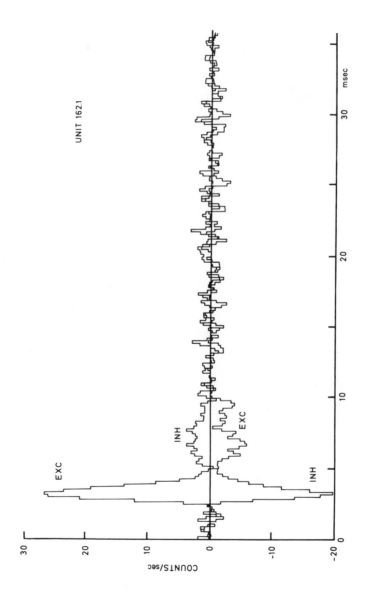

UNIT 162.1

280

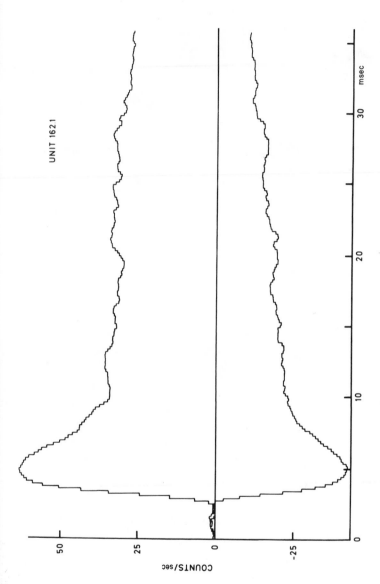

FIGURE 4.12. Estimates of the impulse response (upper graph) and step response (lower graph) functions of the response of a single neuron in the cochlear nucleus of a rat, determined in response to two tones, one of which was amplitude modulated with pseudorandom noise. The curves obtained when the tone at CF was modulated gives an initial upward deflection (marked EXC) and those obtained when the inhibitory tone was modulated show an initial downwards deflection. The frequency of one of the tones was equal to the unit's CF (20.1 kHz) and the other to the unit's best inhibitory frequency (BIF, 22.25 kHz). The intensities of the tones were 50 and 45 dB SPL, respectively. The modulation was 22% RMS. The mean discharge rate of the unit was 110 ppsec. (From Møller, 1976a.)

UNIT 162.2

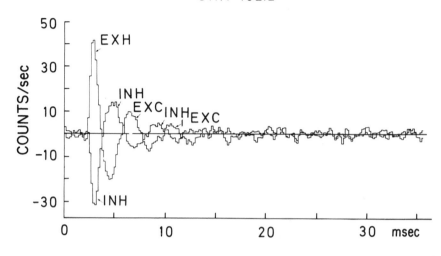

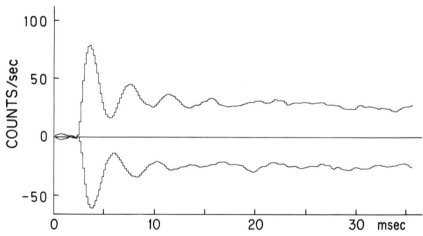

FIGURE 4.13. Graphs similar to those in Figure 4.12 but from a neuron (Type II) showing clearly dampened oscillation in its impulse response to amplitude-modulated sounds. The excitatory tone had a frequency of 18.9 kHz (equal to the unit's CF) and the frequency of the inhibitory tone was 20.9 kHz (equal to the unit's BIF). The intensity of both tones was 75 dB SPL. The average discharge rate of the unit was 220 ppsec. The modulation depth was 22% RMS. (From Møller, 1976a.)

oscillation in the step response of the second type of neuron may be taken as an indication of an inherent regulation of its discharge rate through a feedback control system. The fact that there are recurrent fibers in the cochlear nucleus supports this hypothesis, but no direct evidence showing that a feedback loop exists has been presented. It seems as if both types of neurons enhance fast changes in stimulus intensity in their response patterns. For the Type II neurons, there is a distinct modulation frequency at which the modulation of the discharge rate is greatest. Thus, these Type II neurons exhibit a form of tuning with regard to frequency modulation. The higher the stimulus intensity, the sharper the tuning becomes. A similar tuning has been shown in other sensory neural systems, such as in the eye of the horseshoe crab *Limulus* (Ratliff, Knight, & Graham, 1969).

The Type II units (Figure 4.13) seem to specialize in transmitting changes in sound intensity of a certain rate, keeping the average discharge rate independent of stimulus intensity. The entire dynamic range of the neurons is thus saved for transmitting small, rapid changes in sound intensity.

When the ability of the neurons in the cochlear nucleus to reproduce small changes in the intensity of a sound (modulation) is studied over a large intensity range, it becomes evident that the mean intensity range over which small changes are reproduced, increases when a non-modulated tone is added, if its frequency is located within the inhibitory range of the unit. Stimulation of the inhibitory area of these neurons seems to enhance reproduction of modulation of the discharge rate in response to an amplitude-modulated tone within the unit's excitatory response area. This means that small, rapid changes in the intensity of a tone are reproduced to a greater extent over a larger range of average sound intensities when a background sound (e.g., a tone of slightly different frequency) is present. This is illustrated in Figure 4.14 in which modulation of the period histograms of the discharges of single cochlear nucleus units in response to amplitude-modulated sounds is shown as a function of sound intensity in three different situations, namely, (a) when one single amplitude-modulated tone at CF is presented; (b) when two tones, one inhibitory and one excitatory, are presented simultaneously and the excitatory tone is modulated; and (c) when the inhibitory tone is modulated in the presence of an unmodulated tone at CF.

The degree of modulation of the histograms is shown as a function of the average intensity of stimulus sounds. Modulation of an excitatory tone at CF in the presence of an inhibitory tone results in a larger modula-

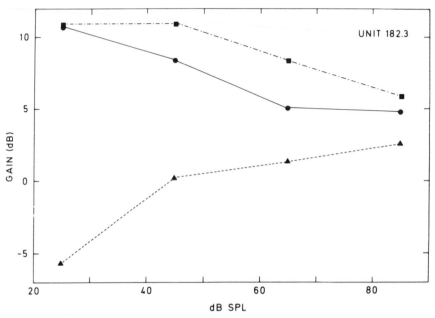

FIGURE 4.14. Relative modulation of the histograms of the responses from a nerve cell in the cochlear nucleus in response to amplitude-modulated sounds as a function of stimulus intensity. The gain is defined as the ratio between the relative modulation of the histograms divided by the modulation of the tone. The three curves illustrate three different stimulus situations: solid line, modulation of a single tone at CF; dashes and dots, stimulation with two tones simultaneously (one inhibitory and one excitatory); dashes, modulation of a single inhibitory tone. (From Møller, 1975c.)

tion of the discharge rate than when the same excitatory tone is modulated but presented without the inhibitory tone being present. Also, the modulation of the discharge rate covers a larger range of intensities when this inhibitory tone is presented than when a single excitatory tone is modulated. Thus, the presence of an unmodulated inhibitory tone seems to facilitate transmission of small changes in intensity (modulation) of a tone. Modulation of the inhibitory tone in the presence of an unmodulated excitatory tone gives rise to less modulation of the histograms than does modulation of the excitatory tone. The finding that reproduction of small amplitude changes is facilitated by the presence of another sound (tone of appropriate frequency) supports the general view that small changes in the intensity of a certain spectral component are preserved in the discharge pattern of single auditory nerve cells in the ascending auditory pathway in the presence of background sounds with a different spectral composition.

Three main features seem to emerge from the experimental results regarding coding of amplitude-modulated tones in the discharge pattern of single nerve cells in the cochlear nucleus:

1. There is specificity with regard to modulation frequency (i.e., modulation with a certain frequency is reproduced to a greater extent than modulation with other frequencies). This is more pronounced in some neurons than in others and two distinct groups may be discerned with regard to being "tuned" to a particular modulation frequency.

2. Amplitude modulation of a tone is reproduced in the discharge pattern of these neurons over a range of average intensities of the tone by far exceeding the range where the average discharge rate is a function of the average sound level. In fact, amplitude modulation is reproduced over a range of intensities that is comparable to the psychoacoustic range of hearing with regard to sound level.

3. The modulation of the discharge rate in response to an amplitude-modulated tone at a unit's CF, becomes larger when an unmodulated tone with frequency and intensity within the unit's inhibitory area is added. This is particularly pronounced at high sound levels. A sound within the unit's inhibitory area thus increases the unit's response to small changes in the amplitude of a tone within the unit's excitatory response area.

CODING OF AMPLITUDE MODULATION IN PRIMARY AUDITORY FIBERS

The responses of single fibers in the auditory nerve to amplitude-modulated sounds have been less extensively studied than those of units of the cochlear nucleus. Comparison between responses of auditory nerve fibers provides valuable information about the transformation between the auditory nerve and the cochlear nucleus. An example of such a comparison between the modulation transfer function of a typical cochlear nucleus unit and a fiber in the auditory nerve is shown in Figure 4.15. The response of auditory nerve fibers seems to reveal a modulation

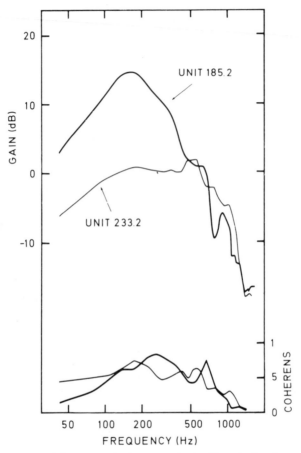

FIGURE 4.15. Modulation transfer functions obtained in a single primary auditory nerve fiber (thin line) and a cell in the cochlear nucleus (heavy line). (From Møller, 1976b.)

transfer function of a lowpass-filter type (Møller, 1976b). Also it follows from available data that the dynamic range of primary fibers, with regard to reproducing the modulation of amplitude-modulated tones, is smaller than that in the cochlear nucleus (Møller, 1976b; Smith & Brachman, 1980). Thus, the firing pattern of cochlear nerve fibers seems to contain information about amplitude modulation only within a relatively narrow range of sound intensities around threshold.

Coding of Natural Sounds and
Neurophysiological Basis of Speech Perception

Compared with simple sounds, we know very little about how natural sounds such as speech sounds are processed in the auditory system. The important features of the sound waves are not understood in detail. It has been hypothesized that there is specificity for the animal's own call sounds in the auditory systems of different animal species.

Some experimental work lends support to the hypothesis that even relatively peripheral neurons show a preference for the animal's own call sounds or sounds to which the animal is known to respond naturally. Thus, it has been shown that the nerve fibers in the frog's auditory nerve have a lower threshold for sounds with spectral distributions that are similar to those of the animal's own call sounds. That would imply that more or less significant anatomical differences exist between the auditory systems of different animals. However, no substantial proof for this hypothesis has been obtained. It is essentially only the frequency range of hearing that has been shown to differ significantly from one mammalian species to another.

The coding of more complex sounds in higher animals, such as the echolocation sounds produced by the flying bat, has been studied recently by recording from single nerve cells in the inferior colliculus (Suga *et al.*, 1979b). It was found that bats respond preferentially to these natural sounds when simpler sounds and natural sounds were presented.

In a series of experiments, Newman and Symmes (1979) studied the responses of single nerve cells in the auditory cortex of the squirrel monkey to the calls of the animal. They related the responses of specific segments of the calls.

Before any conclusions may be drawn from these experiments, however, more experimental work must be done using a greater variety of sounds in a greater variety of animal species. It is naturally of great interest to consider the analysis of speech sounds in the auditory nervous system. Since it is not possible to record from single nerve cells in man, such studies must be based on animal experiments and psychoacoustic experiments in man.

Presently, we base our concept of how sounds such as speech sounds are processed in the auditory system on neurophysiological data obtained in animals and on human psychoacoustic results. Neurophysiological data are sparse and largely incomplete. Furthermore, available results are restricted to coding in single nerve elements. The conclusions that may be drawn from these experiments are hampered by our lack of knowledge about the principles of decoding of the neural discharge pattern. We do not know whether it is possible to draw conclusions about neural processing in one animal species on the basis of results obtained in another species. It may be that different mammals have relatively similar processing of sounds, at least in the ascending auditory pathway up to the auditory cortex, despite the fact that one species may not behaviorally respond to a sound that another responds to.

Kallert (1974) studied the relationship between the responses of single neurons in the medial geniculate and speech sounds. He used trained, conscious cats and recorded from single nerve cells with special telemetric methods. It was found by other investigators that the phonemic content of speech could be distinguished in the recorded multiunit response (Kallert, 1974). Examples of such recordings from single units in the medial geniculate body in Figure 4.16 show how the unit responds to different components of speech sounds. The results shown in Figure 4.16 were obtained in a conscious, freely moving cat using telemetry. The units that responded selectively to speech sounds did not respond to simpler sounds like steady pure tones or frequency-modulated tones. Thus, there seem to be neurons that detect different phonemes and then respond only to certain vowels, certain consonants, or transients of speech sounds. Since complex sounds elicit responses in different neurons on the precortical level, it seems that the prerequisites for decoding speech exist at that level.

Caspary, Rupert, and Moushegian (1977) found that neurons in the cochlear nucleus of the kangaroo rat respond to vowel sounds in a way that indicates a certain specificity with regard to different vowels. In their experiments, such sounds gave rise to a greater diversity of responses than did simpler sounds such as tone bursts.

The great emphasis on spectral properties of speech in connection with spectral analysis in the auditory system may be related to the fact that thus far speech has been described in physical terms, and spectral analysis has played a fundamental role in such description. Speech is most commonly described as a spectogram that shows the distribution of energy of a frequency in various time segments. A spectrogram, or sonogram, shows how the energy is distributed over frequencies and how that pattern changes with time.

Studies of the neurophysiological basis of speech perception greatly

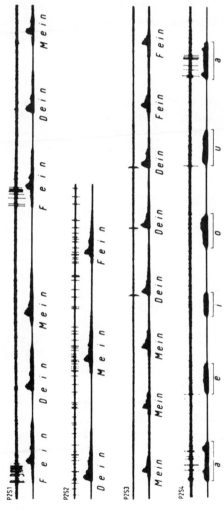

FIGURE 4.16. The response from a single nerve cell in the medial geniculate body of an unanesthetized cat in response to speech sounds. (From Kallert, 1974.)

emphasized the frequency discrimination of the neural activity in the auditory periphery. The studies have usually been designed to test hypotheses regarding how formant frequencies may be resolved by the auditory periphery on the basis of the spectrum analysis performed in the ear. It was assumed that the formant structure of vowels forms the basis for auditory discrimination of vowels. The assumption was based on the finding that different vowels differ in their formant frequencies. However, results of neurophysiological studies on spectral filtering in the auditory periphery suggest that perhaps the spectral cues are *not* extracted and used by the auditory system in recognition of vowels. These experiments instead indicate that the temporal pattern of the vowels is very important. Results were obtained using a method developed by Pfeiffer and Kim (1975) where the responses to the same or a few tonal stimuli were recorded from a large number of primary auditory nerve fibers in the same animal (Pfeiffer & Kim, 1975a; see p. 208). Using this method, but using vowels instead of tones, Sachs and Young (1979) revealed that the formant structures of steady-state vowels were only reproduced in the firing patterns of single auditory nerve fibers at low sound levels. The formant structure of vowels was not seen in the distribution of the discharge rate over a population of auditory nerve fibers above 60 dB SPL. This was assumed to be due to saturation, which prevented further increase in firing rate, and to two-tone suppression. These results indicate that the characterisitics of vowel sounds probably cannot be discriminated by spectral analysis of complex sounds.

Later studies, however, showed that phase-locking (synchrony) of the responses of single auditory nerve fibers to the formant frequencies is maintained over a larger range of sound intensities. Subsequently, the synchrony to nonformant harmonics is suppressed more as the sound intensity is increased (Sachs & Young, 1980).

From this, it may seem that the temporal coding of formant frequencies is more likely to be the way information about formant frequencies is transmitted to higher nervous centers. The possibility that spectral filtering may play a lesser part in the coding of complex sounds is also supported by results of experiments on the coding of spectral distribution of noise sounds in the discharge patterns of single auditory nerve fibers described earlier. These studies also show that the temporal pattern seems to be a more likely carrier of frequency information than the spectral cue. It must be pointed out again, however, that we do not know how either temporal or spatial (spectral) information about such complex sounds as vowels is decoded or otherwise utilized by higher nervous centers. As more results are accumulated, it becomes obvious that it is *not* possible to predict the response pattern of single neurons to complex sounds on the basis of knowledge about their response to simpler sounds.

A number of questions, such as how the peripheral part of the auditory system analyzes sound and how various sound patterns are coded in the discharge pattern in single auditory nerve fibers, have been answered as a result of the great advancements in the ability to record activity of single auditory nerve fibers and the development of new methods for analysis of the recorded discharge pattern. These new results have also given rise to many new questions and have cast doubt upon earlier, well-established conceptions. At a glance, it may, in fact, seem as if we know less about the function of the peripheral part of the auditory system today than we thought we knew some years ago.

It may be worth considering whether or not the various qualities studied in psychoacoustic experiments, such as difference limen for pitch and loudness and a few other attributes of *simple sounds*, are used in perception of such complex sounds as speech. Sounds used in the psychoacoustic experiments that produce these various qualities of sensation are very simple sounds and qualitatively different from the most important sounds for human communication—namely, speech sounds. In other words, there is no doubt that the auditory system can process a certain, experimental sound so that a particular sensation of pitch, for example, is produced in a well-trained, well-motivated human subject. However, do we know if the same processing abilities of the auditory nervous system are used when discriminating natural sounds, such as speech, under everyday conditions? Pitch and loudness sensations may be special qualities that are found only for special, simple sounds. It is not obvious that the sensation of pitch produced by a pure tone or a periodic sound has any relevance in perception and discrimination of such complex sounds as speech sounds. Nor is it directly obvious that other manifestly psychoacoustic attributes of simple sounds, such as loudness, play an important role in the discrimination of speech sounds. Discrimination of speech sounds may not be directly based either on spectral analysis or on periodicity detection.

Signal Processing in the Auditory Nervous System in Summary

The best *short* description of *signal processing* in the auditory nervous system may be that it is a system which enhances changes. These changes may be changes in overall intensity, changes in intensity in certain frequency bands, changes in the frequency of simple sounds, or changes in the spectra of complex sounds.

That description is naturally an oversimplification, and as knowledge about signal transformation in the various nuclei of the ascending auditory pathway accumulates, it becomes clear that the processing of complex sounds is extremely complicated. There is an indication in these results that different types of information are gradually channeled to specialized groups of neurons as the signals proceed toward higher brain levels. Different qualities of a complex sound are thus assumed to be carried to the cortex in separate channels. Whereas a single neuron at the peripheral level can carry information in its firing pattern about many different types of stimuli, in more centrally located nuclei, the individual neurons tend to specialize so that a certain neuron only responds to a certain type of stimulus.

It must also be pointed out that our knowledge about the signal transformation in the auditory system is essentially based on one single type of measurement—namely recordings from single nerve cells. There are a number of other electrical phenomena of single auditory nerve cells that may serve to transmit information, and in addition, there are several chemical processes in nerve cells that may play important roles in transmitting information between nerve cells. Our total lack of knowledge about the principles of decoding information in the nervous system makes it impossible to evaluate the functional importance of a certain electrical or chemical activity in a single neuron. It is also of great importance to note that in most investigations only one nerve cell is studied at a time. We do not know if activity in one nerve cell is important in the discrimination of different types of sensory stimuli. It may be true that it is the activity in groups of nerve cells that is of importance for sound discrimination. It may also be true that the decoding process in the central nervous system is based on the activity or distribution of activity among a number of neurons having certain interrelationships. It seems unlikely that the central nervous system recognizes the discharge pattern in a certain nerve cell. It is more likely that the central nervous system looks for a pattern of responses over populations of nerve cells or differences in response between various populations of nerve cells. The possibility of studying how large populations of nerve cells respond to different types of stimuli is greatly limited. Gross electrode recordings may provide information in that direction, but are generally difficult to interpret. Among other things, this is due to the fact that it is not known which neurons are contributing to the electrical activity recorded and what type of electrical activity in the neuron contributes to the recorded potentials picked up by gross electrodes. Again, lack of knowledge about the principles of decoding makes such speculations or hypotheses of little value in increasing our understanding of the function of the sensory nervous system.

Most work on coding in the auditory nervous system thus concerns coding of particular features in the discharge pattern of single neurons. However, if one asks the simple question: To which sounds will a certain neuron respond? it becomes apparent that there is an increased specialization as one moves from the periphery toward the auditory cortex.

The only known difference in the response patterns of primary auditory nerve fibers is that they respond within specific ranges of intensity and frequency. The discharge pattern contains all information about the temporal pattern of the sound that is utilized in discrimination of the sound. Thus, the only pronounced specificity at the periphery is with regard to frequency or spectrum. Neurons at peripheral levels of the ascending auditory system show specificity to qualities of the sound such as change in amplitude and frequency. Thus, studies of the response pattern of neurons in the cochlear nucleus show that the discharge rate of most of these neurons is modulated with the envelope of an amplitude-modulated tone or noise to a much larger extent than would have been expected on the basis of their response to steady-state (unmodulated) tones or noise. Other neurons seem to integrate energy over a long time and do not respond to fast changes. These neurons may be candidates for transmission of information about overall intensity of a sound (loudness). Neurons in the cochlear nucleus also respond differently to sweep tones (i.e., tones, the frequency of which is varied rapidly over a large range of frequencies—several octaves) than do auditory nerve fibers. The response to such sounds of the cochlear nucleus units cannot be predicted on the basis of their response to steady-state tones or tones with slowly varying frequency. Within a certain range of rates of change, the frequency range of response becomes narrower and more nerve impulses are evoked within a narrow frequency range around the unit's CF than is the case for tones with slowly varying frequency. Some neurons in the cochlear nucleus will only respond to transient sound and show a frequency selectivity that is related to the periodicity of the transients, and not related to the spectrum of the sound. These neurons will not respond to sustained sounds at all. Others respond to different and special features of a sound. Little is known about the response of neurons in the superior olive, except that their responses are specific to stimulation which is ipsilateral, contralateral, or bilateral. The neurons in the superior olive also respond selectively to the frequencies of steady-state tones in much the same way as do neurons in the cochlear nucleus.

The responses of neurons in the inferior colliculus have been studied only briefly and no systematic studies on coding of complex sounds are available. However, the results that have been published seem to indicate that a further specialization occurs in these nuclei. Thus, some neurons in the inferior colliculus will respond exclusively to amplitude-modulated tones, whereas other neurons prefer frequency-modulated sounds (Rees & Møller, 1982). This is specialization, compared to responses of single nerve cells in the cochlear nucleus where the response to frequency-modulated sounds can be explained on the basis of the frequency modulation being converted into amplitude modulation by the frequency selective properties of the neurons. Neurons in the cochlear nucleus will respond to both amplitude and frequency-modulated sounds, whereas many neurons of the inferior colliculus prefer either frequency or amplitude-modulated sounds.

Single nerve cells of the medial geniculate will respond to sound according to more complex criteria. However, too few studies have been performed to give any general description of the response pattern in the nuclei of the inferior colliculus and the medial geniculate.

In the auditory cortex of bats, there are groups of nerve cells that respond to such complicated features of the sound as are related to ranging and directional cues of their echolocating sounds. Some cells respond selectively to features of the call sounds of the particular species (Suga, 1981).

The discharge pattern of single cortical cells is in most cases relatively stereotyped. The cells do not seem to code the temporal pattern of a sound to any great extent. It is also interesting to note that some neurons in the cerebral cortex only respond when a sound has a certain combination of harmonic spectral components. Already in the cochlear nucleus some cells seem to respond over a larger range of intensities when a sound contains a combination of spectral components, in contrast to their response to a pure tone which contains only one spectral component.

When specialization reaches a degree where nerve cells only respond to one particular type of information, the specificity is usually known as *feature detection*. Such neurons are assumed to extract one neural code out of many. Such feature detectors are assumed to activate other parts of the central nervous system selectively when a particular type of sound reaches the ear. Research has shown the existence of such feature detectors for the natural cries of certain animals (Symmes & Newman, 1981). It has been speculated that feature detector neurons exist in the human auditory system and it has been hypothesized that there may be phonemal detectors for speech.

Recently, the tonotopic organization in the auditory cortex has been viewed with new interest and more elaborate tonotopic maps of the cortexes of cats and monkeys have been made on the basis of single cell recordings (Reale & Imig, 1980). It is obvious that there is orderly organization throughout the entire auditory pathway, including the primary auditory cortex (Merzenich & Kaas, 1980). It seems that the orderly tonotopic organization is maintained even in nuclei that receive input from several other (lower) nuclei. An example of this is the inferior colliculus, which receives its input from the lateral lemniscus. Different inputs all seem to project onto the same areas of the inferior colliculus with regard to frequency. Thus, the part of the nucleus that best responds to a particular frequency may receive its frequency-specific input from several nuclei.

The functional importance of tonotopic organization in the various nuclei and particularly in the cortex is obscure, as is the functional importance of the coding of frequency and other information in the discharge patterns of single neurons. There is a fundamental difference between the spatial organization in the visual and somatosensory nervous systems and the tontopic organization in the auditory system. Whereas the visual and somatosensory cortexes (as well as the nuclei in their ascending pathways) are organized two-dimensionally, consistent with the organization of the primary receptors in the retina and the skin, the auditory cortex and relay nuclei show a projection of the basilar membrane, which lacks the dimension of width. The frequency representation therefore takes the form of one or more lines in the auditory cortex and the auditory nucleus, whereas the visual field is represented by a two-dimensional plane on the visual cortex. It is of interest then, to consider what the dimension perpendicular to the projection of the basilar membrane represents. Do these nerve cells represent qualities of a sound other than frequency, or are they just duplicates of neurons responding essentially in a similar way? It might be expected that the separation of different types of information in different neurons described earlier, and seen in the cochlear nucleus and the inferior colliculus, would be reflected in a topical organization with regard to features other than frequency. Examples of such spatial organization of more complex features in the cortex have been demonstrated experimentally, only recently. The results of studies seem to support the assumption that qualities of a sound other than frequency are represented spatially on the cortex (Suga et al., 1979a; Merzenich, & Kaas, 1980; Suga, 1981).

We refer to the work of Suga (1981), who showed that different characteristics of echolocation in the bat such as distance, velocity of a target, etc. are represented in neurons located at separate places on the primary auditory cortex. In a series of elegant experiments, Suga showed that neurons that specialize in such features as ranging are organized in an orderly fashion in the auditory cortex of the animals. Such aggregates of neurons are arranged along axes which contain systematic information-bearing elements. It is particularly interesting that Suga has been able to show that certain relatively complex combinations of spectral components are necessary for certain groups of neurons to be activated. Earlier (Suga et al., 1979a), these investigators found that certain neurons in the cortex of the bat only responded to specific combinations of harmonics of the echolocating sound. It may be that there are similar neurons in other higher mammals that only respond to the particular combinations of spectral components that characterize, for example, vowel sounds.

Results of these experiments indicate that neurons with similar patterns of response to complex sounds are grouped together in the various nuclei and the cortex. Thus, although tonotopic organization in the auditory nervous system has been a well-established concept, it may represent only one feature (frequency selectivity) of the acoustic stimulus that is represented in an orderly way in the nervous system. Other studies have shown other types of specialization. Thus Syka et al. (1981) found that binaural interactions are organized in groups in the inferior colliculus within the same isofrequency area. Merzenich and his colleagues have shown that there is a clear organization of the auditory cortex with regard to binaural interaction (Middelbrooks, Dykes, & Merzenich, 1980; Merzenich & Kaas, 1980).

This binaural information is represented in stripes parallel to frequency projections on the cerebral cortex. These results are other examples of other properties than the frequency of a sound stimulus being represented spatially. It has been clearly shown in these experiments that neurons are not only tuned to specific frequencies of pure tones but are also tuned to other features of a sound. It seems as if the response of a certain group of neurons to a particular sound may depend on the presence of another sound. One might say that sounds may be "permissive."

We do not yet know whether decoding of information is based on the response pattern of groups of neurons or whether it is related to a particular spatial pattern of responding neurons. It may be inferred from results of research that instead of feature detectors (i.e., neurons that are tuned to special and complex properties of a stimulus) particular spatial patterns of response to different characteristics of complex sounds exist.

Finally, we must consider the relevance of these findings. Is it an anatomical difference in the auditory nervous system that causes one mammalian species to respond to certain types of sound that others do not? Or do all mammals have essentially the same ability to respond to sound, but do not do so because of an inhibition either in the ascending auditory pathway or somewhere between the cortex and those other parts of the brain where discrimination takes place? That is, are we equally able to use sound in echolocation as does the flying bat, or are the auditory systems of bats and man different to such a degree that we do not have the ability to use sound in that way? A similar question would be: Can the cat discriminate speech sounds, or is there anything specific in the organization of our own auditory system that is a necessary prerequisite for speech discrimination?

These questions are extremely difficult to answer, but are of great importance when drawing conclusions about the function of our own auditory system on the basis of experimental results obtained in animals.

References

Altman, J. A., Bechtereva, N. N., Radionova, E. A., Shmigidina, G. N., & Syka, J. Electrical responses of the auditory area of the cerebellar cortex to acoustic stimulation. *Experimental Brain Research*, 1976, *26*, 285–298.

Caspary, D. M., Rupert, A. L., & Moushegian, G. Neuronal coding of vowel sounds in the cochlear nuclei. *Experimental Neurobiology*, 1977, *54*, 414–431.

Evans, E. F., & Wilson, J. P. Frequency selectivity of the cochlea. In A. R. Møller (Ed.), *Basic mechanisms in hearing*. New York: Academic Press, 1973.

Gilbert, A. G., & Pickles, J. O. Responses of auditory nerve fibers in the guinea pig to noise bands of different widths. *Hearing Research*, 1980, *2*, 327–333.

Granit, R. The case for relevance in sensory motor physiology. *Trends in Neurosciences*, 1978, *1*, 17–18.

Greenwood, D. D., & Goldberg, J. M. Responses of neurons in the cochlear nuclei to variations in noise bandwidth and to tone-noise combinations. *Journal of the Acoustical Society of America*, 1970, *47*, 1022–1040.

Greenwood, D. D., & Maruyama, N. Excitatory and inhibitory response areas of auditory neurons in the cochlear nucleus. *Journal of Neurophysiology*, 1965, *28*, 863–892.

Kallert, S. *Telemetrische Mikroelektrodencurtersuchungen am Corpus geniculatum mediale der wachen Katze.* Dissertation Erlangen-Nürnberg, 1974.

Marmarelis, P. Z., & Marmarelis, V. Z. *Analysis of physiological systems.* New York: Plenum Press, 1978.

Merzenich, M. M., & Kaas, J. H. Principles of organization of sensory-perceptual systems in mammals. In J. M. Sprague & A. N. Epstein (Eds.), *Progress in psychobiology and physiological psychology* (Vol. 9). New York: Academic Press, 1980.

Middlebrooks, J. C., Dykes, R. W., & Merzenich, M. M. Binaural response-specific bands in primary auditory cortex (AI) of the cat: Topographical organization orthogonal to isofrequency contours. *Brain Research,* 1980, *181,* 31–48.

Møller, A. R. Unit responses in the cochlear nucleus of the rat to sweep tones. *Acta Physiologica Scandinavica,* 1969, *76,* 503–512.

Møller, A. R. Unit responses in the cochlear nucleus of the rat to noise and tones. *Acta Physiologica Scandinavica,* 1970, *78,* 289–298.

Møller, A. R. Unit responses in the rat cochlear nucleus to tones of rapidly varying frequency and amplitude. *Acta Physiologica Scandinavica,* 1971, *81,* 540–556.

Møller, A. R. Coding of sounds in lower levels of the auditory system. *Quarterly Review of Biophysics,* 1972, *5,* 59–155.

Møller, A. R. Responses of units in the cochlear nucleus to sinusoidally amplitude modulated tones. *Experimental Neurology,* 1974, *45,* 104–117.a

Møller, A. R. Use of stochastic signals in evaluation of the dynamic properties of a neuronal system. *Scandinavian Journal of Rehabilitation Medicine,* 1974, *3,* 37–44.b

Møller, A. R. Coding of sounds with rapidly varying spectrum in the cochlear nucleus. *Journal of the Acoustical Society of America,* 1974, *55,* 631–640.c

Møller, A. R. Dynamic properties of excitation and inhibition in the cochlear nucleus. *Acta Physiologica Scandinavica,* 1975, *93,* 442–454.

Møller, A. R. Dynamic properties of the responses of single neurons in the cochlear nucleus of the rat. *Journal of Physiology* (London), 1976, *259,* 63–82.a

Møller, A. R. Dynamic properties of primary auditory fibers compared with cells in the cochlear nucleus. *Acta Physiologica Scandinavica,* 1976, *98,* 157–167.b

Newman, J. D., & Symmes, D. Feature detection in squirrel monkey cortex. *Experimental Brain Research,* 1979, *2,* 140–145.

O'Neill, W. E., & Suga, N. Target range sensitive neurons in the auditory cortex of the mustache bat. *Science,* 1979, *203,* 69–73.

Pfeiffer, R. R., & Kim, D. O. Cochlear nerve fiber responses: Distribution along cochlear pattern. *Journal of the Acoustical Society of America,* 1975, *58,* 867–869.

Ratliff, F., Knight, B. W., & Graham, N. On tuning and amplification by lateral inhibition. *Proceedings of the National Academy of Sciences U.S.A.,* 1969, *62,* 733–740.

Reale, R. A., & Imig, T. J. Tonotopic organization in auditory cortex of the cat. *Journal of Comparative Neurology,* 1980, *192,* 265–291.

Rees, A., & Møller, A. R. The responses of neurons in the inferior colliculus of the rat to AM & FM tones. *Hearing Research,* 1982 (in press).

Ruggero, M. A. Response to noise of auditory nerve fibers in the squirrel monkey. *Journal of Neurophysiology,* 1973, *36,* 569–587.

Sachs, M. B., & Young, E. D. Encoding of steady-state vowels in the auditory nerve: Representation in terms of discharge rate. *Journal of the Acoustical Society of America,* 1979, *66,* 470–479.

Sachs, M. B., & Young, E. D. Effects of nonlinearities on speech encoding in the auditory nerve. *Journal of the Acoustical Society of America,* 1980, *68,* 858–875.

Siebert, W. M. Stimulus transformations in the peripheral auditory system. In P. A. Kolers & M. Eden (Eds.), *Recognizing patterns.* Cambridge, Massachusetts: MIT Press, 1968.

Sinex, D. G., & Geisler, C. D. Auditory-nerve fiber responses to frequency-modulated tones. *Hearing Research,* 1981, *4,* 127–148.

Smith, R. L., & Brachman, M. L. Response modulation of auditory nerve fibers by AM-stimuli: Effects of average intensity. *Hearing Research,* 1980, *2,* 123–133.

Spoendlin, H. The innervation of the cochlear receptor. In A. R. Møller (Ed.), *Basic mechanism in hearing.* New York: Academic Press, 1973.

Suga, N. *Cortical representation of biosonar information in the mustached bat.* The 28th International Conference Physiological Sciences, Publishing House of Hungarian Academy of Sciences, 1981.

Suga, N., & O'Neill, W. E. Neural axis representing target range in the auditory cortex of the mustache bat. *Science,* 1979, *206,* 351–353.

Suga, N., O'Neill, W. E., & Manabe, T. Harmonic sensitive neurons in the auditory cortex of the mustache bat. *Science,* 1979, *203,* 270–274.

Syka, J., Druga, R., Popelár, J., & Kalinova, B. Functional organization of the inferior colliculus. In J. Syka & L. Aitkin (Eds.), *Neuronal mechanisms of hearing.* New York: Plenum Press, 1981.

Symmes, D. On the use of natural stimuli in neurophysiological studies of audition. *Hearing Research,* 1981, *4,* 203–214.

Index

A

Acoustic admittance, *see* Acoustic impedance

Acoustic impedance, 20–39
  of the cochlea, 30
  of the ear, 20–31, 34–39
    individual variation, 24
    middle ear muscles, contraction of, 41–47
    middle ear cavities, 34

Acoustic middle ear muscles, *see* Stapedius muscle; Tensor tympani muscle

Acoustic middle ear reflex, 39–66, 109
  anatomy, 39
  drugs, 55–58
  hearing one's own voice, 63
  pure tones, 47–48
  reflex arc, 40
  sensitivity, 47–51
  sensitivity to complex sounds, 48–51

temporal characteristics, 51
temporary threshold shift, 64
transmission, 58

Acoustic middle ear transmission, 34–39, 58–63
  air pressure, 34–39
  cavities, 34–38
  muscle contraction, 58–63

Adaptation, 159–166
Admittance, *see* Acoustic impedance
Afferent fibers, 71–74
Amplitude modulated sounds, 277–286
  auditory nerve, 286
  cochlear nucleus, 277–285

Ascending auditory pathway, 108–116

Auditory cortex, *see* Cortex, auditory

Auditory nerve, 85, 109, 118–123, 286
  fibers, 136–162
    rate intensity curves, 145–146